Digital Government

Digital Government

Digital Government

TECHNOLOGY

AND PUBLIC SECTOR

PERFORMANCE

Darrell M. West

PRINCETON UNIVERSITY PRESS

PRINCETON AND OXFORD

ISBN: 0-691-12182-6

Library of Congress Cataloging-in-Publication Data

West, Darrell M., 1954–
    Digital government : technology and public sector performance / Darrell M. West.
       p.   cm.
    Includes index.
    ISBN 0-691-12182-6 (cl: alk. paper)
    1. Internet in public administration—United States.   2. Administrative agencies—United
States—Data processing—Evaluation.   3. Political participation—United States—
Computer network resources.   4. Democracy.   I. Title.

JK468.A8W44   2005
352.3′8′02854678—dc22   2004057274

British Library Cataloging-in-Publication Data is available

This book has been composed in Electra

Printed on acid-free paper. ∞

pup.princeton.edu

Printed in the United States of America

10 9 8 7 6 5 4 3 2 1

To the faculty of the Political Science Department at Miami University of Ohio, for giving me a terrific undergraduate education

Contents

Contents

List of Tables

Preface

THE SUBJECT OF DIGITAL GOVERNMENT first came to my attention several years ago while searching public sector websites for information about online services. At that time, it struck me that many government websites were difficult to use and lacked a standard design for visitor navigation. Despite optimistic rhetoric regarding the so-called e-government revolution, it was not clear to what extent the Internet was transforming public sector performance or democracy. Given the fact that electronic government ("e-government") still was in its infancy, it seemed an appropriate time to track government websites as well as how citizens were thinking about e-government.

This book looks at how e-government has developed and how the Internet compares with historical examples of technological change. Every time a new technology has emerged, there have been grandiose claims about its impact on society and politics. From the printing press in the 1500s to inventions such as the telegraph, telephone, radio, and television, technical innovations often are said to produce a major transformation. At their creation, for example, both the telegraph and television were cited as new devices that would speed communication and alter the relationship between citizens and government.

It therefore is no surprise that the emergence of the Internet and other digital technology has led to speculation about their longer-term social and political consequences. Are they transforming government? How are they altering public sector performance? What are their ramifications for the way democracy functions? As in the past, scholars have attempted to determine what these latest new technologies mean for our political system.

In this research, I examine the extent to which e-government has transformed the public sector, what factors dictate the extent of change, and the ramifications of e-government for public performance and democracy. Basically, I suggest that e-government falls more within a model of incremental change than of transformation. Using data on the content of government websites, citizen and bureaucrat attitudes toward e-government, expenditure data, case studies, an email responsiveness test, and aggregate multivariate analysis, I argue that like many past technologies, the Internet's effect has been mediated by a variety of political forces. Factors such as group conflict, bureaucratic setting, and budget scarcity have slowed the rate of innovation and made it difficult for e-government to take advantage of the Internet's revolutionary potential. This limits the transformational scope of the Internet and slows the diffusion of information technology.

A number of people helped me collect data on and think about the ramifications of e-government. Melissa Driscoll Nicholaus provided invaluable research assistance on this project. I also would like to thank the legion of students who put in long hours documenting the content of government websites: Kristine Hutchinson, Todd Auwarter, Nicole Scimone, and Melissa Iachan during summer 2000; Benjamin Clark, Kim O'Keefe, Julia Fischer-Mackey, Sheryl Shapiro, Chris Walther, Shih-Chieh Su, Ebru Bekyel, and Mariam Ayad during summer 2001; Dylan Brown, Bill Heil, Jason Holman, Aiko Wakau, Julie Petralia, Joshua Loh, Ones Umut, Irina Paley, Marilia Ribeiro, and Yen-Ling Chang during summer 2002; and Erica Dreisbach, Joanne Chiu, Emily Boness, Carrie Bersak, Adam Deitch, Vanessa Wellbury, Toby Stein, Ones Umut, Irina Paley, Fredi Chango, and Yen-Ling Chang during summer 2003. In addition, Kim O'Keefe and Robert Mitzner helped gather materials for the case study of online tax filing.

I appreciate the assistance of the Council for Excellence in Government (especially President Patricia McGinnis) for making available the raw e-government data of their national public opinion surveys and the surveys of bureaucrats conducted by Peter Hart and Robert Teeter. These data were very valuable for studying citizen and administrator reactions to e-government.

Funding for this project was provided by the Taubman Center for Public Policy at Brown University, the university's graduate school in the form of a Salomon Grant, and the office of the Dean of the College through its Undergraduate Teaching and Research Assistantship program. I am also grateful for funding provided by the Benton Foundation (to write a policy briefing report), the Washington Resource consulting company (for the online tax case study), and the World Markets Research Center (for the first year of the global e-government study).

Earlier portions of this research were published by the *Public Administration Review* in a January/February 2004 article entitled "E-Government and the Transformation of Service Delivery and Citizen Attitudes," Volume 64, number 1, pp. 15–27. I am grateful to the editors for permission to reprint some of those results in this volume.

I appreciated the chance to try out portions of the arguments presented in this book to the following audiences: the American Political Science Association annual meeting in San Francisco (2001); the St. Petersburg, Russia, E-Government Conference (2002); St. Petersburg Technical University (2002); the Karelia National Library (2002); the Beirut, Lebanon, E-Government Conference (2002); the American University of Beirut (2002); Notre Dame University in Lebanon (2002); the North-West Academy of Public Administration in St. Petersburg, Russia (2003); and information technology officials in Taipei, Taiwan (2003), and Tokyo, Japan (2004). I received a number of helpful comments from people at these events.

Chuck Myers, political science editor at Princeton University Press, deserves a big thank you for his insightful commentary on this book. He made a number of suggestions that improved the content and organization of the argument. The same was true for the book's reviewers. They helped me refine my argument and evidence presented in this volume. Jonathan Munk did an excellent job editing the book. Debbie Tegarden was prompt and thorough in organizing the book's production. I am grateful for the assistance of each of these individuals.

Digital Government

Digital Government

CHAPTER 1

Scope, Causes, and Consequences of Electronic Government

THIS BOOK LOOKS at the phenomenon of electronic government, that is, public sector use of the Internet and other digital devices to deliver services, information, and democracy itself. Although personal computers have been around for several decades, recent advances in networking, video imaging, and graphics interfacing have allowed governments to develop websites that contain a variety of online materials. As more and more people take advantage of these features, digital government is supplanting traditional means of access based on personal visits, phone calls, and mail delivery.

In Indiana, for example, citizens can register their vehicles and order subscriptions to government databases online. California allows people to personalize websites depending on whether they are tourists, students, state employees, businesses, or state residents. Arizona and Michigan have been innovators in online voting. At the national level, Americans can access private companies through the Internal Revenue Service (IRS) website that will file tax forms for them electronically.

Governments around the world have created websites that facilitate tourism, citizen complaints, and business investment. Tourists can book hotels through the government websites of many Caribbean and Pacific island countries. In Australia, citizens can register government complaints through agency websites. Nations such as Bulgaria, the Netherlands, and the Czech Republic are attracting overseas investors through their websites.

But despite the prevalence of these online options, there are three unanswered questions that form the heart of this research. First, how much are the Internet and other digital delivery systems transforming the public sector? Second, what determines the speed and breadth of e-government adoption? Third, what are the consequences of digital technology for public sector performance, the political process, and democracy?

One of the problems in deciphering these topics has been a bifurcation of e-government research into detailed case studies on the one hand and highly abstract theoretical treatises on the other hand. There are a variety of case study publications that focus on particular agencies or jurisdictions. These are very useful, but hard to generalize to a larger universe of cases. At the other end of the spectrum are highly theoretical treatments of e-government that either glorify technology or study it at abstract levels. With either brand of theorizing, it is a challenge for practitioners and policymakers to figure out how to use this information to improve their performance.

My goal in this book is to bridge the worlds of theory and practice. E-government is a field in which practitioners and theorists need to address one another and share their respective insights. It is vitally important that we have clear conceptual frameworks for the analysis of e-government. It also is crucial that these frameworks rest on empirical analysis that actually shows what is happening and what problems need to be addressed.

Consistent with this objective, I look at the scope, causes, and consequences of digital government. Using multiple methods (case studies, content analysis, public and bureaucrat opinion survey data, an email responsiveness test, and aggregate multivariate analysis), I present a conceptual model in which factors such as organizational setting, budget resources, group conflict, and political leadership set the parameters on the speed and breadth of technological change. I also explore the consequences of e-government for service delivery, the needs of special populations such as the disabled and non-native speakers, bureaucrat attitudes and behavior, citizen trust in government, public sector responsiveness, and overall political dynamics. With the exception of impact on bureaucrats, most of these topics have attracted little scholarly attention.

After examining a variety of data sources, I argue that e-government falls more within models of limited than transformational change. There are a variety of forces that restrict the ability of policymakers to make effective use of new technology. Although digital breakthroughs offer the potential of revolutionary change, social, political, and economic factors constrain the scope of transformation and prevent government officials from realizing the full benefits of the Internet.

THE SCOPE OF E-GOVERNMENT: HOW MUCH CHANGE?

The origin of the Internet dates back to 1969 when a United States Defense Department project spawned ARPANET, a digital system connecting computers in different geographical locations. These connections allowed scientists at fifteen different universities and American defense officials to exchange information and post notes at common computer spaces that could be viewed simultaneously by interested parties.[1]

Unlike telephones, which required communicating parties to be on the line at the same time for the transmission of material, ARPANET allowed people to send information even if the other person was not at the other end to receive the transmission. Scientists could transmit emails or access bulletin boards, and thereby see ideas on which others were working without physically being in the same location. This type of asynchronous communication proved very popular with scientists and members of the defense establishment.

It was not until 1991, though, with the formation of interfaces with the World Wide Web that the Internet was created as a means of communication among the general public. The Web integrated text, images, and sound, and therefore facilitated the instantaneous communication of several modes of information. Unlike past electronic systems that required extensive technical knowledge or involved specialized programming, it was simple to use.

Within a few years, government agencies discovered that the Internet was a useful way to communicate with citizens, businesses, and other agencies. Information and services could be put online and made available to a wide variety of people.[2] Right now, the Internet remains the most popular e-government delivery system. Eighty-one percent of federal e-government initiatives are delivered that way, with the remainder coming in the forms of kiosks, telephones, email, bulletin boards, and wireless networks.[3]

Citizens enjoy the convenience of digital communications as a means to contact government officials with problems they are having. A July 2003 Pew Internet Survey found that of those Americans who had contacted the government in the past year, 26 percent called on the phone, 18 percent visited the government website, 12 percent went to the agency in person, 10 percent sent a letter, and 10 percent relied upon email.[4]

By the turn of the twenty-first century, some governmental units were employing the Internet for direct democracy. When faced with a controversy over the repainting of a prominent Baltimore expressway bridge (a local artist wanted rust red-brown while the mayor preferred Kelly green), Mayor Martin O'Malley turned to an online plebiscite at the City Hall website. Voters were asked to choose which color the Howard Street Bridge should be painted. Over five thousand people cast ballots, and the mayor's choice lost on a 52 to 48 percent vote. In his concession speech, O'Malley said "I submit to the will of the people. To paraphrase Jefferson, I fear for my countrymen, that they may receive the bridge colors they deserve."[5]

Government websites at various levels generate considerable traffic. As shown in table 1-1, the U.S. Internal Revenue Service is the most frequently visited individual site with 6.7 million monthly visitors, followed by Fed-World, a federal government portal with 4.6 million visitors, the U.S. Treasury (3.63 million), and NASA, with 3.33 million visitors.

Unlike traditional bricks-and-mortar agencies that are hierarchical, linear, and one-way in their communications style, digital delivery systems are non-hierarchical, nonlinear, interactive, and available twenty-four hours a day, seven days a week. The nonhierarchical character of Internet delivery frees citizens to seek information at their own convenience. The interactive aspects of e-government allow both citizens and bureaucrats to send as well as receive

TABLE 1-1
Most Frequently Visited Government Websites in February 2001

IRS (irs.gov)	6.7 million
FedWorld (fedworld.gov)	4.6 million
U.S. Treasury (ustreas.gov)	3.63 million
NASA (nasa.gov)	3.33 million
California state government (ca.gov)	2.89 million
Natl Institute of Health (nih.gov)	2.49 million
US Education Dept (ed.gov)	2.46 million
Natl Oceanic and Atmospheric Admin (noaa.gov)	2.2 million
US Geological Survey (usgs.gov)	1.43 million
US Navy (navy.mil)	1.41 million
Library of Congress (loc.gov)	1.17 million
White House (whitehouse.gov)	1.09 million
Social Security Admin (ssa.gov)	1.06 million
Texas state government (state.tx.us)	1.06 million

Source: Jupiter Media Metrix, March 13, 2001, press release
Note: Figures represent number of unique agency visitors for the month of February 2001.

information. Convenience is probably the strongest selling point for e-government as ordinary citizens love the ability to access public information and order services twenty-four hours a day, not just when a particular government agency happens to be open.[6]

The fundamental nature of these advantages has led some to predict the Internet will transform government.[7] By facilitating two-way interaction, electronic governance has been hailed as a way to improve service delivery and responsiveness to citizens. Stephen Goldsmith, President George W. Bush's Special Advisor for Faith-Based and Community Initiatives, says "electronic government will not only break down boundaries and reduce transaction costs between citizens and their governments but between levels of government as well."[8]

Jeffrey Seifert and Matthew Bonham argue digital government has the potential to transform governmental efficiency, transparency, citizen trust, and political participation in transitional democracies. Using examples from Asia and Eastern Europe, these authors suggest that with proper political leadership, the power of the Internet can be harnessed for major system change.[9]

In making these claims, proponents suggest that the pace of Internet change is consistent with the classic model of large-scale transformation. System transformation is defined as a "complete change in character, condition," or "epochal breakthroughs."[10] In this perspective, change is rapid and abrupt, and visible to social observers. Often spurred either by scientific breakthroughs or economic improvements that facilitate the availability of the new technology, large-scale change produces revolutions in individual behavior and organizational activities.[11]

For example, the novel aspects of digital technology in the public sector led Reed Hundt, former chairman of the Federal Communications Commission, to conclude "the central lesson of technology in our time is this: The Internet Changes Everything. The lesson applies to the economy, education, community, individualism, and . . . democracy."[12] Similar statements have been made about the impact of the new information age on the political process. Writers such as Dennis Thompson and Bruce Bimber have suggested some aspects of interactive technologies bring about institutional change because they weaken the factionalization that plagues democratic political systems. New technologies enhance communication by overcoming geographical distance, promoting ideological variety, opening citizens to more diverse viewpoints, and encouraging deliberation.[13] These benefits give the Internet unusually great promise as a tool for democracy.

Others have written about the potential of the Internet to recast bureaucracy. Jane Fountain has discussed the ways in which information technology (IT) alters the capacity and control features of traditional bureaucracies. IT, she notes, has the potential "to substantially redistribute power, functional responsibilities, and control within and across federal agencies and between the public and private sectors."[14] By encouraging bureaucrats to work together and develop cross-agency "portals," websites that integrate information and service offerings, e-government offers the prospect of considerable change in how the public sector functions. Indeed, Fountain cites estimates demonstrating "cost performance ratios to be declining at a rate of 20–30 percent a year."[15]

Not all technological innovation, however, leads to large-scale transformation. An alternative model stresses incrementalism. First proposed in 1959 by Charles Lindblom in regard to organizational decision making, this kind of change is characterized as a "muddling through" process.[16] In looking at how organizations make choices, Lindblom asked whether change was rational and dictated in key respects by economic trade-offs or was it rather a political process characterized by small-scale shifts constrained by budgetary and institutional processes?

In the world of government, Lindblom suggested, politics dominates and organizations are more likely to muddle through decisions and rely on small-scale change. Political dynamics affect the way in which decisions get made. It is not always the most rational decision that emerges based on costs and benefits. Rather, choices get made based on who is best organized, strongest politically, or in control of the bureaucratic structure. The political character of public sector decision making limits the speed of change and how quickly new technologies get incorporated into the governmental process.

Taking off from this insight, Aaron Wildavsky and others generalized Lindblom's process model to policy outputs.[17] Government policies typically evolve through small-scale steps, not large-scale transformations, he argued. The best predictor of next year's budget is this year's budget. Change generally takes place in small increments, which leads to gradual change over time. Abrupt and dramatic revolutions in political behavior are rare. Evolution, not revolution, is the more common norm.

IT research in the 1970s and 1980s found considerable evidence of incremental change in government organizations. Work by Kenneth Kraemer, John King, and William Dutton demonstrated that American governments at every level were slow to adopt new technologies.[18] Rather than being an impetus toward transformation, computer technologies were not used to produce fundamental change.

There are a number of reasons why political change tends to be small in scale and gradual. Government actions are mediated by a range of factors: institutional arrangements, budget scarcity, group conflict, cultural norms, and prevailing patterns of social and political behavior, each of which restricts the ability of technology to transform society and politics. The fact that governments are divided into competing agencies and jurisdictions limits the ability of policymakers to get bureaucrats to work together promoting technological innovation. Budget considerations prevent government offices from placing services online and using technology for democratic outreach. Cultural norms and patterns of individual behavior affect the manner in which technology is used by citizens and policymakers.

In addition, the political process is characterized by intense group conflict over resources. With systems that are open and permeable, groups organize easily and make demands on the political system. Given the fact that financial resources are limited and institutions in which decisions are made are fragmented and decentralized, it is difficult to produce large-scale changes even with the benefit of new technologies.

With many government planners emphasizing a vision of electronic governance that is technocratic and service-oriented rather than a tool for grassroots empowerment, system-level transformation has been slow to develop. Regardless of the type of political system, many government officials are conservative when it comes to change. Rather than rushing to embrace new technology, major political and economic interests slow the pace of technical innovation until they can figure out how to make sure their own vested interests are well-protected. This keeps the danger from new technology as low as possible, and forces technology to accommodate existing power structures rather than the other way around.

These kinds of political constraints are so widespread that Richard Davis, Michael Margolis and David Resnick, Andrew Chadwick, and Christopher May predict in the long run that Internet technology will *not* transform democracy. If anything, technology reinforces existing social and political patterns rather than creating new realities. In regard to technology, Davis notes, "that complex bureaucratic maze also has been duplicated on the Web." Agency websites serve to perpetuate their own mission and do little to enhance responsiveness or citizen participation.[19] Margolis and Resnick argue that "far from revolutionizing the conduct of politics and civic affairs in the real world, we found that the Internet tends to reflect and reinforce the patterns of behavior of that world."[20] Chadwick and May found government websites in the United States, Great Britain, and European Union to be "predominantly non-interactive and non-deliberative," and concluded e-government was not likely to reshape governance.[21]

Transformation and incrementalism are the major poles in the debate over the pace and breadth of technological change. In the debate over the transforming power of new technology, however, it is important to remember that change represents a continuum running from incrementalism to transformation. Change discussions often focus on the end points of this comparison without looking at other types of models. There are lots of ways in which shifts can occur in the middle of the change spectrum. Or to put it differently, there are other models of change between large-scale transformation and small-scale increments.

Secular change is an example of a midlevel model that demonstrates how "constrained change" can unfold. For example, James Quinn develops a model of "logical incrementalism" that suggests significant change can take place within organizations on a step-by-step basis even outside of a revolutionary change model.[22] The cumulative impact of steady, incremental change over a long period of time is major, he suggests.

In the same vein, Fountain's notion of "enacted technology" discusses change that is substantial even if it trails proponents' projections.[23] It is not uncommon for creators of particular technologies to "oversell" the significance of their invention. Many creations are said to be revolutionary in their potential to reshape human life. When the reality falls far short of predictions, however, the resulting change looks underwhelming and is dismissed as insignificant change. Yet Fountain points out that the alterations that emerge in this situation still can be significant even if they fall short of what was predicted by those who created the new technology.

In essence, both of these writers offer an insight into an alternative model of technological change emphasizing gradual, secular change that unfolds slowly but surely over time, and eventually leads to major changes in how organizations function. Revolutions do not have to be quick and abrupt for there to be widespread change.[24] It may take awhile for technical innovations to diffuse throughout a country. In the public sector, there may be a period of bureaucratic in-fighting that impedes the adoption of new technologies. Sometimes, years may pass before the price of helpful inventions drop to the point where it becomes feasible for individuals and organizations to adopt them.

The automobile is a good illustration of a secular change model. It took decades after the placement by German Karl Benz in 1885 of the internal combustion engine on a motorized carriage for the "car" to emerge as a dominant form of transportation that would have enormous social, political, and economic consequences. The first American automobile was built by Charles and Frank Duryea in 1893, but it took them three years to construct thirteen cars. Ransome Eli Olds started the first automobile assembly line in 1901, but his factory burned down before cars could be produced. Henry Ford introduced the Model T in 1908. He built a mass-production line and eventually this paved the way for the manufacturing of millions of cars.[25]

But it was not until decades later that cars began to transform the social and political landscape. In the 1950s and 1960s, the car plus the development of interstate highways encouraged people to move from central cities to the suburbs. This had enormous social and political ramifications. Jobs and residences moved to the suburbs, which drained population and economic vitality from the cities. Political power moved outside of metropolitan areas with these population flows and politicians began to play more to suburban interests. Eventually, central cities went into decline and became bastions of poor people, minorities, and senior citizens with few economic resources.

Ultimately, as with any political issue, it is impossible to know when a particular technological innovation will produce large-scale societal change.[26] Change is a mosaic that unfolds in a kaleidoscope of colors. It never is easy to decipher longer-term impact. But this does not mean that judgments must await the passage of decades after the technical invention was developed. If that were the case, it would be impossible to deal with the deleterious effects of change until it is too late. Given the uncertainty of long-term change, it makes sense in the short run to focus on the causes and consequences of new practices, and how substantial the alteration is in individual and organizational activities.

The virtue of studying short-term change is that it provides hints about longer-term shifts. By providing policymakers with benchmarks for evaluating how close they come to achieving particular goals, short-term change assessments help direct the future impact of particular technologies. In that way, then, policymakers can see where things are headed and what midcourse alterations are necessary to head off negative consequences.

Stages of E-Government: From Billboards and Service Delivery to Interactive Democracy

There are four general stages of e-government development that distinguish where government agencies are on the road to transformation: (1) the billboard stage, (2) the partial service-delivery stage, (3) the portal stage with fully executable and integrated service delivery, and (4) interactive democracy with public outreach and accountability-enhancing features.

This categorization does not mean all government websites go through these exact steps or that they undertake them in a linear order. It is clear from looking at agency websites that there is a variety of ways in which e-government has evolved. Based on our research of looking at thousands of websites, however, this sequence appears to be a prevalent course of action in many jurisdictions. The commonality of this model therefore allows researchers to distinguish agency progress based on how far along they are in incorporating various website features.

In the first stage, officials treat government websites much in the way highway billboards are used, that is, static mechanisms to display information. They post reports and publications, and offer databases for viewing by visitors. There is little opportunity for citizen interaction and no chance for two-way communications between citizens and officials. Visitors can read government reports, see the text of proposed legislation, and check to find out who works in specific offices.

Even today, some government offices, such as those of the U.S. Congress, remain stuck in the billboard stage. A study of member websites undertaken by Congress Online at George Washington University found most representatives and senators are using their sites as "promotional tools." Rather than having services or interactive technologies, legislators are employing their websites to post "press releases, descriptions of the Member's accomplishments and photos of the Member at events."[27]

The static nature of a billboard approach limits a visitor's ability to use interactive technologies. Citizens can see information, but not alter it to their own ends. Government websites utilizing this approach offer the advantage of access to information, but do not allow citizens to search the site, send feedback, or order government services. Without the ability to "engage" a government website, citizens cannot take advantage of the technology's capacity for two-way communications or personalize the website to their own specific interests.[28]

Due to these limitations, some government agencies have moved to a second stage, that of incorporating information search features and partial service delivery into the website. In this phase, citizens can access, sort, and search informational databases. Government websites start to place some services online, although the services offered tend to be sporadic and limited to a few areas.

This stage represents an advance over the billboard approach, but there are limits to what citizens can do online. In this situation, most government agencies are slow to incorporate truly interactive features onto their websites. Citizens are not able to "personalize" their website or engage in conversation with public officials. There is little way to take full advantage of the power of digital technologies.

The third stage features "one-stop" government portals with fully executable and integrated online services. This phase offers considerable convenience to visitors. The entire city, state, or nation has one place where all agencies can be accessed. This improves citizen ability to find information and order services. Agency sites are integrated with one another and a range of fully executable services are available to citizens and businesses. Officials show that they pay attention to privacy and security concerns on the part of the general public by posting policies online. No longer are websites static and presentational, but dynamic and interactive. By incorporating advanced features on government websites, citizens gain control over information and service delivery. Visitors can register to receive updates and newsletters, as well as other material that is useful to them.

The limiting factor of this stage, however, is that it is characterized more by a service-delivery mentality than by a vision of transforming democracy.[29] Government websites generally have been slow to take advantage of "democracy-enhancing" technologies that would improve responsiveness to citizens or help the public hold leaders accountable for governmental actions. Public planners are more apt to want to get new services online than seek to extend democracy to disenfranchised citizens. There is little interest in providing opportunities for government feedback and public participation in decision making.

This stage ignores the central virtue of the Internet: its ability to enhance the performance of democratic institutions and improve the functioning of democracy. Technology is available, though not widely implemented, for citizens to convey preferences to government personnel, participate in agency decisions, and improve the functioning of democratic political systems. Few of these attributes have been incorporated into the public sector, however, because government officials emphasize a model of e-government based on service delivery as opposed to system transformation. The public sector is less apt to think of the Internet as a tool for fundamental institutional change than for the delivery of particular services to businesses and the middle class.

While these two visions are not necessarily mutually exclusive, they do represent different emphases, and lead to major variations in e-government priorities. The more service delivery dominates e-government thinking, the less likely government websites are to incorporate interactive features that help the site achieve the full potential of democratic governance. Rather than devising opportunities for participation and representation, many government websites emphasize service delivery to current Internet users.

It is at the fourth stage—interactive democracy with public outreach and accountability measures—that government websites move to a goal of system-wide political transformation. In addition to having integrated and fully executable online services, these kinds of government sites offer options for website personalization (i.e., customizing for someone's own particular interests) and push technology (i.e., providing emails or electronic subscriptions that

TABLE 1-2
E-Government Stages and Models of Technological Change

Billboards	Partial Service Delivery	Portal Stage with Fully Executable and Integrated Services	Interactive Democracy
Key qualities include reports, publications, and databases, but no services or interactive features.	This stage allows visitors to search websites and order a few limited services. There are few privacy or security statements and no means to personalize site.	Site has online services, integrated across agencies. Substantial concern with privacy and security. Some means to obtain electronic updates.	Lots of online services and interactive features. Site features accountability-enhancing features and technologies for public feedback and deliberation.
Incremental Change			
Secular Change			
Transformational Change			

Source: Author compilation

provide automatic updates on issues or areas people care about). These features help citizens customize information delivery and take advantage of the interactive and two-way-communications strengths of the Internet. Through these and other interactive features, visitors can avail themselves of a host of sophisticated technologies designed to boost democratic responsiveness and leadership accountability.

As pointed out by Thomas Beierle, it takes a long time for agencies to become interested in incorporating principles based on online political participation into their mission. A study he conducted of the Environmental Protection Agency relied on a two-week online discussion entitled "Democracy Online: An Evaluation of the National Dialogue on Public Involvement with EPA Decisions." It brought together 1,166 people for an electronic discussion in the form of a bulletin board. According to his analysis, "most people reported being satisfied with the process." The major complaint registered was that "experts" tended to dominate the online discussion, thereby discouraging participation by ordinary citizens.[30] But this effort to incorporate citizen participation represented a novel use of interactive technology on government websites.

These four stages of e-government provide a rubric by which to gauge the effectiveness of technology and the degree of technological change. As shown in table 1-2, the movement from the billboard stage to interactive democracy represents the clearest evidence of transformational change. Governments that incorporate tools of democratic outreach, interactive elements, privacy and security policies, and accountability-enhancing elements in their websites come closest to fulfilling the revolutionary claims of Internet visionaries.

In contrast, movement from billboards to portals with fully executable and integrated service delivery is consistent with models based on secular change. Here, the changes are significant, but not revolutionary. Government planners are incorporating elements that serve the middle class and make it easier to access public sector services, but are not using the Internet as a tool for system transformation. The vision is technocratic, rather than citizen-empowering.

Meanwhile, evidence of incremental change comes when websites move from the billboard stage to partial service delivery. In this situation, officials are incorporating new technology, but at a slow pace. They are not making much use of interactive features. Their goal is not to transform the political or governmental system, but to add discrete improvements that make it easier to access online information and services.

The Causes of E-Government: What Drives the Speed and Breadth of Technological Change?

In assessing the factors that drive alterations in individual behavior and institutional performance, it is tempting to reify technology and emphasize a technology-driven perspective. Rudi Volti, for example, notes how "technological determinism" pervades some analysis of change.[31] According to this approach, technology itself determines change simply by the force of the new invention. If the telegraph speeds up information transmission across the entire country, then that aspect of the telegraph becomes the reason why newspapers adopt a more national perspective. Or if television provides visual images, then the rise of telegenic politicians is attributed to the visual dimension of television.

This approach, however, ignores the fact that the longer-term impact of technology is mediated by organizational setting, political dynamics, media coverage, and budget realities.[32] An organizational approach posits that the pace and breadth of change is affected by factors such as the nature of work routines within bureaucratic agencies and the degree to which the organization is open to change. These factors have enormous consequences for the speed of diffusion of technology, people's receptivity to using new technology, and the extent to which inventions transform society and politics.[33]

As pointed out by Fountain, sometimes the bureaucracy is a barrier in technological innovation because most new creations represent a change in the status quo. Each new innovation forces bureaucrats to alter routines, develop new working relationships, and sacrifice autonomy. Bureaucrats can slow or speed the diffusion of innovation by placing barriers in the path of new ideas. Even technologies with a demonstrated record of efficiency and effectiveness will not be adopted unless government officials decide that invention should be implemented.

In addition, the revolutionary potential of new technology is affected by political dynamics.[34] Because of their need to provide universal access, government organizations suffer from what is called the "two systems" problem.[35] This dilemma arises when agencies seeking to innovate technologically must maintain parallel systems of information and service delivery (face-to-face, telephone, and mail) at the same time they are building electronic interfaces. Interest group pressures dictate that public sector agencies cannot shut down government offices or stop answering the phone when they create email or Internet delivery systems because many people lack digital access.

The degree of political conflict has ramifications for how the two systems issue is handled and the manner in which new technology is integrated into the public sector. As Lindblom and Wildavsky have pointed out, the incorporation of technology into government inherently is a political process. Interest groups compete to make government decisions as favorable to their interests as possible. This could happen either through labor-management negotiations, use of government contracts, or lobbying by outside organizations.[36]

Groups that are well-organized typically are in a stronger position to gain favorable decisions from government because they provide votes, money, and/or volunteers crucial to the survival of office-holders. Regardless of whether a politician works in the legislative or executive branch, group demands and resources are important in determining whether particular innovations are adopted by government agencies. It is not merely a neutral or technical process in which good technologies rise on their merits. Group claims shape which technologies are adopted and at what rate.

Financial resources and budget conditions are other factors that drive the pace of governmental change. Technology requires up-front investment, and the relative scarcity or abundance of budget support makes a huge difference to the ability of government agencies to innovate. The more difficulty bureaucrats have financing new technologies, the more difficult it will be for those innovations to be adopted and produce large-scale change. During periods of economic prosperity, governments are in a stronger position to innovate than during times of budget deficits. Resources set the broad parameters under which government officials negotiate group demands and navigate bureaucratic settings in the public sector.

Media coverage is important to the dissemination of new technology because it affects both how people think about technology and their receptivity to change. Reporting that is positive about technology encourages people to be favorable to new creations. Inventions that have negative side-effects or get mired in partisan scandals and contracting controversies are going to diffuse much more slowly through the public sector.

Political leadership matters because strong cues from elected officials or top administrators encourage public sector organizations to speed or slow down the adoption of new technology. A governor who wants his or her state to be in the forefront of e-government can overcome bureaucratic intransigence, find resources that facilitate innovation, and resolve group conflict that slows down the pace of diffusion. Leadership that is open to innovation or sees an electoral payoff from adopting technology can make a tremendous difference in how technology gets integrated into the public sector.

As discussed at greater length in chapter 2, it is important to look at organizational, fiscal, and political factors that drive the pace of change. The bureaucratic setting in which individuals make decisions about new technology, the nature of group interests and conflict, and political leadership all affect how quickly technology is introduced into government agencies. Unless these forces are understood, it will be impossible to determine whether new creations fall within models of technological change based on transformation, secular change, or incremental alterations.

THE CONSEQUENCES OF E-GOVERNMENT: HOW THE INTERNET AFFECTS THE PUBLIC SECTOR, POLITICS, AND DEMOCRACY

Technology has something of a checkered past with respect to its long-term impact on society.[37] For example, inventions such as nuclear fission have been used both militarily and pacifically. Sometimes, even peacetime applications of nuclear energy generation have created devastating environmental consequences. The ability of technology to produce both positive and negative consequences means observers have to be alert to the wide variety of changes that emerge from inventions. It is not enough merely to assess what is happening; one must also ascertain how desirable particular shifts are. Do they increase the effectiveness, efficiency, and responsiveness of a system? This question is particularly relevant to the subject of the Internet, which already has created considerable controversy concerning security, privacy, and content.

Given the complexity of technological change, it never is easy to determine the ultimate impact of new technologies on society and government.[38] As will be made clear in this study, assessing the long-term consequences of technological change is a challenging task. Sometimes, decades must pass before the ultimate impact of inventions becomes clear. Change does not unfold clearly or uniformly. Rather, there are a number of different avenues by which technology emerges.

When the Internet first appeared, people embraced it as the perfect tool for personal liberation. Due to its decentralized character and capacity for two-way interaction, proponents sold it as a nirvana that would give citizens complete control over their information requirements. Rather than having the government or big media companies control information dissemination, the ordinary person would be empowered to make his or her own choices.

Yet in an era of unwanted spam, viruses, computer hackers, and security breakdowns, people are reassessing the societal benefits of electronic technology. Rather than a tool for liberation and empowerment, the Internet has been plagued by behavior that invades personal privacy, causes large-scale inconvenience, and threatens confidential material.

In 2001 and 2002, for example, the Computer Security Institute estimated that losses from computer viruses alone totaled nearly $50 million, not to mention the countless hours of personal aggravation suffered by computer users.[39] Researchers claim that 10.4 million spam emails are sent every minute of the day around the world.[40] Not surprisingly, in the face of this spam onslaught, a 2003 Pew Internet Survey found that 70 percent of email users complain that being online has become "annoying or unpleasant" and half report they are "less trusting of email" than before.[41]

In thinking about the impact of e-government on the public sector, there are a number of particular aspects that need to be assessed: information availability, serving special populations, online service delivery, democratic responsiveness, democracy enhancement through interactive features, and citizen trust in government. The most basic question concerns the availability of information and whether information is accessible to people with special needs, such as non-native speakers and the physically disabled. Further, citizens enjoy the convenience of accessing information online and not having to call, visit, or mail requests for government documents. Being able to go online and view government reports and databases helps citizens understand what the public sector is doing and how government officials are performing their basic duties.

Service delivery is another aspect of e-government evaluation. What services are online and how well are they functioning? Are they the types of services citizens find useful and that make their lives easier? For many citizens, being able to access services online is one of the most desirable aspects of e-government. It saves time and effort, and is much more convenient than having to visit a government agency in person and wait in line for an extended period.

Democratic responsiveness refers to the degree to which e-government improves a system's capacity to respond to ordinary people. The shared advantage of search engines, email contact information, and interactive features is the empowerment of the common person in his or her dealings with the public sector. Search engines give people control over information by allowing them to search for what they want, as opposed to what a government official may want to show them. Email contact points give citizens a means for notifying government officials when there is a problem or the person wants to lodge a complaint or make a suggestion.

Democracy enhancement refers to the ability of technology to improve democratic performance beyond responsiveness. This could range from simple things such as placing audio or visual materials online (such as broadcasts of hearings or speeches) to more interactive mechanisms that allow citizens to vote, make comments on proposed government rules, or personalize websites to their particular interests. Any of these mechanisms is helpful because it employs technology to tailor usage to the needs and interests of those accessing the site.

Finally, there is the question of how the Internet affects overall political dynamics.[42] How does it affect the relationship between politicians, bureaucrats, and information specialists? Will e-government alter the balance of power between these individuals and improve citizen attitudes toward government? In the long run, proponents have argued that e-government will improve service delivery at lower cost, and thereby transform citizen attitudes toward the public sector. It is important to examine such claims to see if there is improved government performance and whether these improvements lead to shifts in how people see the public sector.

DATA AND METHODS

To get a full picture of how e-government is progressing, I relied on a variety of data sources from content analysis and survey analysis to case studies and aggregate multivariate analysis. I undertook a detailed content analyses of 17,077 American city, American state, American federal, and foreign government websites from 2000 to 2003 (see table 1-3). Our research team analyzed 5,005 city websites, 6,146 state websites, 274 federal websites, and 5,651 global websites.

The city sites were from the 70 largest American metropolitan areas, as defined by the U.S. Census Bureau. We looked at an average of 22 sites per city in 2001, 22 in 2002, and 28 in 2003. The state analysis was based on sites in each of the fifty states (an average of 34 websites for each individual state in 2000, 32 sites per state in 2001, 25 sites per state in 2002, and 32 sites per state in 2003). The global sites were from the 198 countries around the world. We attempted to get up to 20 sites per country, but found many small countries had only one or a few agencies on their website. So we analyzed as many sites as we could find for these countries, and ended up with an average of 12 sites per country in 2001, 6 in 2002, and 11 in 2003.

In terms of case selection, this analysis included sites from each branch of government in the various levels of government. Among the sites analyzed were those developed by court offices, legislatures, Congress, state and national officials, major cabinets and departments, and state and federal agencies serving crucial functions of government, such as health, human services,

TABLE 1-3
Number of Government Websites Studied in Content Analysis, 2000–2003

	2000	2001	2002	2003	Total
City	—	1,506	1,567	1,933	5,005
State	1,716	1,621	1,206	1,603	6,146
Federal	97	58	59	60	274
Global	—	2,288	1,197	2,166	5,651
Total	**1,813**	**5,473**	**4,029**	**5,762**	**17,077**

Source: Author's e-government content analysis database

taxation, education, corrections, economic development, administration, natural resources, transportation, elections, and business regulation. Websites for obscure state boards and commissions were excluded from the study.

For the global sites, we looked at those of executive offices (such as a president, prime minister, ruler, party leader, or royalty), legislative offices (such as Congress, Parliament, and various people's assemblies), judicial offices (such as major national courts), cabinet offices, and major agencies serving crucial functions of government, such as health, human services, taxation, education, interior, economic development, administration, natural resources, foreign affairs, foreign investment, transportation, military, tourism, and business regulation.

Recognizing that there is no agreement on appropriate benchmarks of what constitutes a good or effective government website, we developed our own analysis based upon online features that have been judged important by citizens in market research and opinion surveys: a page with contact information, links to publications and databases, access to services, privacy and security, usability by populations with special needs such as the disabled and non-English speakers, and readability level. Each website was evaluated for the presence or absence of more than two dozen different features at the point in time we visited that site (see appendix I for details on coding these websites).[43] We also conducted detailed tests of the readability level of government websites (using the Flesch-Kincaid test) and their accessibility to the disabled (using the automated "Bobby" software provided at http://bobby.watchfire .com). The entire site for every agency was studied to provide a complete picture of its contents.

I also analyzed the raw data of a national public opinion survey conducted August 14–16, 2000, with 1,003 randomly sampled adults across the United States. This telephone survey had a margin of error of plus or minus 3.5 percent and was undertaken by the polling firm of Peter Hart/Robert Teeter of Washington, D.C., on behalf of the Council for Excellence in Government. This survey sample (as well as follow-up surveys in 2001 and 2003) was developed using random-digit-dialing sampling techniques and included an over-sample of 200 frequent Internet users. Data were weighted in accordance with the demographic composition of the United States population. Seventy-nine questions were included, such as items measuring the use of government websites, evaluations of e-government (including ease of finding sites, overall rating, and past and future positive impact), views about government and political activity (trust in government, confidence in government, views about government effectiveness, and measures of political activity), and common political and demographic controls (sex, age, race, income, education, and party identification) (see appendix I for question wording and order).

This public poll was followed with another survey in 2001 that looked at whether opinions had shifted in light of the September 11 terrorist attacks. The 2001 survey was sponsored by the Council for Excellence in Government and completed by Hart/Teeter. It was based on interviews with 961 adults nationwide during November 2001. Several questions in 2001 were repeated from the 2000 survey in order to facilitate comparability. The 2001 survey had a margin of error of plus or minus 3.5 percent.[44]

In 2003, the Council for Excellence in Government sponsored another survey focusing on public opinion towards e-government. This research, also undertaken by Hart/Teeter, included interviews with 1,023 adults across the country in February 2003. It had a margin of error of plus or minus 3.1 percent.[45]

To see how bureaucrats felt about e-government, I studied a Council for Excellence in Government survey of bureaucrats undertaken August 10–18, 2000, plus follow-up surveys in November 2001 and February 2003. Each survey was completed by Hart/Teeter and investigated how directors and managers felt about electronic governance. The 2000 survey included interviews with 150 local, state, and federal government administrators; the 2001 survey was based on the views of 400 administrators; and the 2003 project interviewed 408 administrators. Each of these surveys utilized many of the same questions as on the public survey, such as overall assessments of e-government, what aspects of the public sector are being affected, and concerns about e-government. The 2000 bureaucrat survey had a margin of error of plus or minus 7 percent, while the 2001 and 2003 surveys had margins of error of plus or minus 5 percent.

In order to examine responsiveness to citizen requests, our research team sent an email in 2000 to four offices in each state—the governor, the legislature, the top state court, and the major human services agency—as well as to all major federal agencies. We undertook similar tests in 2001, 2002, and 2003. The message asked a simple question: "I am trying to find out when your agency is open. Could you let me know the official hours your office is open? Thanks for your help." Email responses were recorded based on the number of business days it took each agency to respond. These results help us analyze the degree of e-government responsiveness to citizen requests and how that responsiveness changed over time.

I gathered budget data outlining state government expenditures on information technology for fiscal years 1998, 1999, and 2000. This information was compiled by the National Association of State Information Resource Executives (now known as NASCIO, or the National Association of State Chief Information Offices). These data show the percentage of the state budget devoted to information technology, and how those figures changed between 1998 and 2000. Of the fifty states surveyed by NASCIO, 46 percent (23 states) provided IT budget figures. The jurisdictions responding included both large states (Texas, Ohio, Michigan, and Pennsylvania) and small states, as well as a mix of "innovating" and "following" states.

I undertook a detailed case study of online tax filing at the federal and state level to get a sense of progress on the most widely used online service. Although there is considerable variation across the fifty states as well as between the approaches at the state and federal levels, the analysis sheds light on how this service was put online, the role of outside contractors, the cost of online service delivery, and the reactions of the general public.

Finally, I completed an aggregate multivariate analysis of e-government performance at the state (and national) level to explain why some states (and nations) have made greater progress than others. Using various indicators of e-government activity, I developed state-level (and national) indicators and modeled their ability to distinguish jurisdictions that have made progress from those that have not. Among the topics examined were the impact of wealth, organization, and democratization on overall e-government performance, number of online services, quality of privacy policies, and readability level.

PLAN OF THE STUDY

The outline of this study is as follows. Chapter 2 examines the bureaucratic, fiscal, and political setting in which e-government has taken place. By studying these types of explanatory factors, I investigate how these determinants of technological change are affecting the ability of the Internet to improve government service delivery. In general, I find that several of these factors are constraining technological change and slowing the rate of innovation in the public sector.

Chapter 3 studies the content of e-government. Drawing on an analysis of thousands of government websites at various levels (United States city, state, and federal), I show how the rate of change is proceeding and how quickly progress is being made at putting materials online.[46] In addition, I investigate the degree to which special populations such as non-English speakers, those who are not very literate, and the physically disabled are being served by e-government. This analysis demonstrates that e-government is adding features in a manner consistent with an incremental change more so than transformation.

Chapter 4 seeks to explain why some governments have done better at incorporating technology into their websites than have others. Using the various performance indicators presented in chapter 3, I develop aggregate statistical models that explain e-government activities, such as the number of online services, accessibility to the disabled, quality of privacy policies, and website readability. Briefly, I find that factors such as state wealth and legislative professionalism are keys to the development of online government in the American states. There is, however, considerable variation in explanatory factors depending on which aspect of electronic governance is being studied. There is no association between e-government performance and privatization or budget deficits in the states.

Chapter 5 presents a case study of state and federal online tax filing. Because this service generates revenue, it has been very popular with government officials. This case study examines how putting tax filing online has progressed, what problems have emerged, and what this area tells us about the ability of technology to alter people's behavior. By focusing on a particular online service that is being utilized by millions of Americans, I show both the potential of and the limits on the so-called e-government revolution.

Chapter 6 focuses on a particular aspect of e-government: its ability to improve democracy and responsiveness to citizens. These are key normative values in the governmental area. Proponents have claimed that e-government will improve responsiveness and bring citizens closer to political leaders. By looking at the extent of democratic outreach in American e-government (meaning using the interactive aspects of the Internet to reach out to citizens) and how responsive bureaucratic agencies are to citizen requests, I argue that e-government has not dramatically improved interactivity, responsiveness, or public outreach.

Chapter 7 incorporates a citizen perspective on e-government. It focuses on who is going online and what they like and dislike about digital government. Drawing on public surveys, I study the extent to which e-government is afflicted by a digital divide and how this affects the ability of the Internet to transform the public sector. Not surprisingly, there is extensive heterogeneity in terms of reliance on and attitudes toward e-government. Various groups of citizens differ considerably in how much they like electronic governance. There are big differences in usage levels by age and education, and these differences affect the ability of various groups to take advantage of online information and services.

Chapter 8 looks at the ability of e-government to improve citizen trust and confidence in the public sector. For years, Americans have been cynical about the government, feeling that it is inefficient and inept at solving problems. E-government proponents, however, claim the Internet offers the potential to turn around citizen attitudes and improve citizen confidence in government. This section presents evidence showing the degree to which these hopes are being met, and ways in which e-government has the potential to draw citizens and leaders closer together.

Chapter 9 focuses on global e-government. How are other nations coping with electronic governance? How does the public sector in foreign nations compare to that in the United States in terms of e-government initiatives? Are the determinants of foreign e-government similar to America? In what ways do the varying cultures, bureaucratic settings, and political systems affect e-government? This section looks at the 198 nations around the world to see how e-government is progressing and whether factors such as wealth, organization, and democratization are constraining the incorporation of new technology. In addition to a detailed content analysis of foreign government websites, I undertake an aggregate analysis at the national level to determine why some countries have made faster e-government progress than others.

Chapter 10 steps back from the particular findings reported in this study and discusses the relationship between democratization and e-government performance. I show that nondemocratic systems are as likely as democracies to perform well on new technology initiatives. Some authoritarian countries have been successful with digital government because they have top-down political structures and are able to overcome bureaucratic and political intransigence. Most political regimes have features that limit technological change and discourage system transformation. To illustrate this point, I compare the Internet with historic inventions such as the printing press, telegraph, telephone, radio, and television, and argue that the slow rate of organizational change seen with the Internet is similar to that of many past creations. In each era, there have been a variety of political and institutional factors that have constrained the rate of diffusion. I close the volume by discussing what can be done to facilitate technological innovation and make public sector organizations more receptive to change.

CHAPTER 2

Bureaucratic, Fiscal, and Political Contexts

BUREAUCRATS AND POLITICIANS are central to public sector decisions to adopt new technology. They allocate funds and mediate conflict. They decide whether to outsource technological innovation or undertake it in-house. They juggle competing priorities and assemble political coalitions that either facilitate or slow the rate of technology diffusion within specific government agencies.

In recent years, officials have faced major changes that are relevant for their technology decisions and overall capacity for organizational innovation. There have been alterations in American public opinion, the philosophy of government, financial resources, and political dynamics that have major ramifications for electronic governance. These developments affect how public sector officials handle their jobs and the speed with which specific technologies have been integrated into an agency's mission.

In this chapter, I look at the role of bureaucrats and politicians in technological change. Administrators have been portrayed in many bureaucracy writings as resistant to change and more interested in their own autonomy than agency performance as a whole.[1] As a result of the pervasive public mistrust toward government and the introduction of an administrative philosophy known as the New Public Management, however, bureaucrats today are more receptive to technological change than is commonly thought. Many government administrators view the Internet favorably and see it as a force for improving public sector performance. This positive personal predisposition to technology, though, does not reduce the constraining influences of agency fragmentation, limited budget resources, group competition, media coverage, and partisan conflict. A number of these features have slowed the diffusion of technology within government. With these limitations on the pace and breadth of technological change, it has been difficult for e-government to make rapid progress in incorporating new technology into agency operations.

CITIZEN MISTRUST OF GOVERNMENT

Americans today are remarkably cynical about government and its attendant bureaucracy. This is very different from the situation of several decades ago. Public opinion surveys from the 1950s show that about two-thirds of citizens trusted the government in Washington to do what is right. People had confidence that government officials were motivated by general interests and were not dishonest or corrupt in the way they handled public policy. While there were occasional scandals that tested this trust, voters viewed instances of corruption as individual failings, not systemic maladies.

The situation today could not be more different. For a number of years, surveys have demonstrated that about two-thirds of Americans are cynical and do not trust the government to do what is right. Government is thought to be the tool of special interest groups, political leaders are widely considered to be dishonest and corrupt, and government itself is seen as bloated, inefficient, and ineffective at solving basic problems.

Bureaucrats are one of the reasons many Americans no longer trust the public sector. State and federal administrators are viewed as inefficient, ineffective, and unresponsive to external pressures. John Huber, Charles Shipan, and Madelaine Pfahler have discussed how legislators often come into conflict with bureaucrats over the best course of action. When administrators prefer their own policies, politicians exert tight statutory controls to make sure administrators follow the desired path.[2]

In the same vein, Michael Alvarez and John Brehm have analyzed why public attitudes toward the Internal Revenue Service (IRS) have become so negative in the past two decades. They claim "the word 'bureaucracy' is usually considered so negative that virtually every recent major scholarly consideration of the performance of bureaucracy begins with an acknowledgment of the poisonous connotation of the word."[3] One only has to look at recent elections to find a large number of candidates campaigning against faceless bureaucrats and administrators pursuing their own interests.

Confidence in government also has been harmed by inadequate public sector performance. For the past several decades, there have been periods of unsatisfactory performance in areas such as the economy and foreign affairs. For example, in the 1970s, the country was hurt by the combination of high unemployment and high inflation (leading to the term "stagflation").[4] This led not just to poor ratings for individual office-holders, such as then President Jimmy Carter—dissatisfaction with Carter's performance on the economy and in foreign policy led to a general sense that things were not going well in America—but citizen unhappiness with the entire political system.

Additionally, serious mistakes in judgment and the dissemination of outright lies by elected officials led many Americans to conclude that their government was not looking out for collective interests. President Lyndon Johnson's lies about the progress of the war in Vietnam and President Richard Nixon's deceit and obstructionism in the Watergate scandal undermined the fundamental bases of public trust in government.[5] President Ronald Reagan's poor judgment in the Iran-Contra scandal further weakened citizen confidence in government. And, most recently, personal scandals have plagued several prominent politicians, such as former President Bill Clinton and former Speaker of the House Newt Gingrich.[6] These periodic embarrassments have allowed numerous conservative and liberal politicians over the past few decades to campaign against the Washington establishment.

Each of these factors has contributed to a general unhappiness among Americans with how their political system is functioning. Public opinion surveys demonstrate that many people feel that government spends too much money.[7] Members of both major parties have campaigned to cut "wasteful" public sector spending in a variety of policy areas. Even though the Democrat Clinton reclaimed the White House in the 1990s, there was no cessation of anti-Washington rhetoric. Clinton himself sought to move his party to the center and disavow his party's history of big programs, especially after his health care reform plan was rejected. With the quiet acquiescence of liberals in the Democratic party, the country's public philosophy turned away from large government bureaucracies to a "leaner" government featuring greater reliance on privatization and outsourcing.

THE NEW PUBLIC MANAGEMENT

Consistent with the rise of anti-Washington and antigovernment attitudes is an administrative philosophy known as the New Public Management. In the United States and in many other countries around the world, this perspective reflects the triumph of business models in the public sector. Key to this approach is the viewpoint that there needs to be greater emphasis on "results" in the public sector and that market competition and outsourcing to private sector contractors will improve government responsiveness and efficiency. Citizens are thought of as "customers" and government agencies are expected to become more entrepreneurial and innovative in their use of public monies. Rather than have hierarchical and insulated layers of management, the thinking is that the professional civil service should be flattened and middle-level managers empowered to innovate. In the long run, the expectation is that the resulting reorganization of government functions will produce greater efficiencies and better performance.

The ascendance of this philosophy dovetailed with the negative portrait of government bureaucrats that focused on their inefficiency, selfish desire for autonomy and power, and resistance to change of any type. Taking cues from writers such as David Osborne, President Clinton in 1993 initiated a new effort called "reinventing government." This movement aimed to end the model of a hierarchical, top-down bureaucracy that was unresponsive to its clients. As stated by Osborne, "in a world of rapid change, technological revolution, global economic competition demassified markets, an educated work force, demanding customers, and severe fiscal constraints, centralized, top-down monopolies are simply too slow, too unresponsive, and too incapable of change or innovation."[8]

Led by then Vice President Al Gore, this enterprise sought to remake the public sector more along the mold of private businesses. The bureaucracy would be streamlined and the civil service would have more incentives to re-engineer agencies. Some functions would be outsourced to private companies and others would be run through cross-agency task forces. According to this effort, the resulting public sector would be in a stronger position to innovate and respond to external challenges.

Clinton wanted to alter the longstanding impression that the American bureaucracy was beset by inefficiency and ineffectiveness. When he announced the National Performance Review (NPR) in 1993, the president said, "our goal is to make the entire federal government both less expensive and more efficient, and to change the culture of our national bureaucracy away from complacency and entitlement toward initiative and empowerment. We intend to redesign, to reinvent, to reinvigorate the entire national government."[9] NPR members analyzed individual agencies and oversaw "reinvention laboratories" that suggested changes in agency procedures and operations.

With the World Wide Web expanding rapidly in the 1990s, Clinton had the perfect technology to implement his reinvention vision. Public officials looked toward the Internet as a technology that would bring greater efficiency, effectiveness, and responsiveness to government.[10] Writers such as Douglas Holmes incorporated private sector reasoning in his popular book on e-government and argued that government needed to act more like business.[11] It should focus on customers, be results-oriented, and become entrepreneurial in its activities. Indeed, much of the impetus for bureaucratic change during this period came from government using Internet technology to improve performance.

Under Clinton and Gore, the federal government downsized. With the end of the Cold War, the military was reduced in size and many domestic agencies faced tight fiscal restraints. A number of sectors experimented with privatization and outsourcing to private companies in order to achieve greater efficiencies in the public sector. Although labor unions and some advocacy groups were not happy with the Democratic Party's shift to the middle of the political spectrum and its emphasis on reduced growth of government programs, President Clinton continued to emphasize the importance of government revitalization.

When George W. Bush became president in 2001, this reliance on private sector thinking was driven to new heights. Pushing much further than ever envisioned by the Clinton administration, Bush championed himself as the first "CEO" president, reflecting both his political philosophy and his education as a Harvard University MBA. Newspaper articles proclaimed that Bush would lead "the most corporate-oriented administration in American history."[12] In addition to emphasizing private sector ideas, Bush also brought into his cabinet leaders from the corporate world, such as Vice President Dick Cheney, the former head of the energy services company Halliburton.

Although many of his administration's ideas for downsizing the civil service were controversial, terrorist bombings of New York City and Washington, D.C., on September 11, 2001 gave Bush additional impetus and political resources with which to push for dramatic change. In response to this tragedy, Congress created a new department of Homeland Security that would coordinate federal programs dealing with domestic and foreign intelligence-gathering. Unlike past federal agencies, however, none of this department's 170,000 employees were given civil service protections. In addition, half of the remaining federal civil servants outside the area of homeland security found their government jobs privatized to outside contractors and corporations. It was the most dramatic remaking of the federal bureaucracy in more than half a century.

The triumph of this negative view of the traditional government bureaucracy is particularly noteworthy in that it runs contrary to the opinions of some prominent academics. In his pathbreaking book, *Dismantling Democratic States*, Ezra Suleiman challenged market-based thinking and said it was time to rethink the New Public Management.[13] A well-trained professional civil service is crucial to public sector performance, Suleiman claimed, and democratic political systems cannot function effectively without a strong civil administration. The New Public Management erred, he noted, by thinking government could and should act like private business. In reality, the two have very different missions and divergent responsibilities.

These contrasting views of government bureaucrats have attracted considerable attention when it comes to bricks-and-mortar government. A number of scholars have looked at the relationship between bureaucratic performance and the civil service sector.[14] There has, however, been little empirical analysis of bureaucratic attitudes toward the newly emerging prospect of e-government. Much of the debate over the predispositions of bureaucrats in that area has been noticeably devoid of data on actual attitudes. What do government administrators think about e-government? How open are they to technological change? What fears and concerns do they have about its long-term future? And how do their views compare to those of the general public?

BUREAUCRATS AND E-GOVERNMENT: AN EMPIRICAL ANALYSIS

Surveys of government administrators at the local, state, and federal levels in the United States demonstrate how substantially e-government is affecting public sector employees. In August 2000 the Council for Excellence in Government undertook an attitudinal study of 150 local, state, and federal employees about e-government. Among the topics posed were questions regard-

TABLE 2-1
Personal Involvement of Government Administrators in E-Government, 2000–2003

	2000 (%)	2001 (%)	2003 (%)
Very Involved	21	28	29
Fairly Involved	27	29	54
Somewhat Involved	33	28	12
Not Involved	18	14	5
Not Sure	1	1	0

Source: Council for Excellence in Government National Surveys of Government Administrators

ing how personally involved that administrator was in e-government. This survey was repeated with 400 employees in November 2001 and 408 employees in February 2003.

As shown in table 2-1, there has been a major increase in government employees' personal involvement with e-government. In 2000, 48 percent of government administrators reported they were involved in e-government. This percentage rose to 57 percent in 2001 and 83 percent in 2003, reflecting increased levels of e-government activity.

In general, these surveys found high receptivity to Internet technology and extensive optimistism about its ability to improve government functioning among managers and administrators (table 2-2). In 2000, 83 percent said they thought e-government was having a positive effect on the way government operates. This was similar to the 78 percent who felt that way in 2001 and 82 percent who thought so in 2003.

In holding these views, government administrators are much more favorable about e-government than the general public. As shown in table 2-3, bu-

TABLE 2-2
Evaluations of E-Government by Government Administrators, 2000–2003

	2000 (%)	2001 (%)	2003 (%)
Very Positive	48	40	36
Somewhat Positive	35	38	46
Neutral	11	15	11
Somewhat Negative	0	1	0
Very Negative	1	0	0
Not Sure	5	6	7

Source: Council for Excellence in Government National Surveys of Government Administrators

TABLE 2-3
Evaluation of E-Government by Bureaucrats and the General Public, 2000

	Bureaucrats (%)	General Public (%)
Positive	83	35
Neutral	11	22
Negative	1	11
Unsure	5	32

Source: Council for Excellence in Government National Survey of Government Administrators and the General Public, 2000

reaucrats are almost 50 percent more positive about e-government than are citizens in general. The public was more likely to hold negative views or to be unsure about its feelings.

In 2000, many administrators believed e-government would change public sector activities for the better (table 2-4). For example, 91 percent believed e-government would improve people's ability to get information, while 89 percent thought it would help the government offer easy-to-use services. Only 45 percent believed e-government would improve privacy protection and 47 percent thought it would help to keep personal information secure.

When asked what were the most positive things to have emerged from e-government, there were interesting changes since 2000. As shown in table 2-5, there was little change in access to information or the offering of more convenient government services. There was, however, an increase from 17 to

TABLE 2-4
Percentage of Government Administrators Believing E-Government Would Change Various Public Sector Activities for the Better, 2000

Ability to Get Information	91
Ability of Government to Offer Easy-to-Use Services	89
Ability of Government to Offer Convenient Services	87
Coordination across Government Levels	83
Ability of Employee to Do Job Well	83
Ability to Communicate with Elected Representatives	78
Accountability to Citizens	75
Cost-Effectiveness	73
Level of Personal Service from Government Employees	69
Ability to Keep Personal Information Secure	47
Level of Privacy Protections	45

Source: Council for Excellence in Government National Surveys of Government Administrators

TABLE 2-5
Most Positive Things to Emerge from E-Government, in View of Government
Administrators, 2000–2003

	2000 (%)	2001 (%)	2003 (%)
Greater Public Access to Information	34	38	36
More Efficient and Cost-Effective Government	17	16	28
More Convenient Government Services	23	22	21
More Accountable Government to Citizens	19	13	9
Government That Provides More National/Homeland Security	NA	6	4

Source: Council for Excellence in Government National Surveys of Government Administrators

28 percent in administrators believing e-government would produce more efficient and cost-effective government and a decrease from 19 to 9 percent in administrators thinking it would raise government accountability. This suggests government officials are more optimistic about efficiency than accountability improvements resulting from e-government.

The bureaucrat survey also revealed that since 2000 administrators increasingly are worried about the lack of financial resources for e-government (table 2-6). In 2000, only 26 percent cited this as one of the biggest obstacles, but this percentage rose to 44 percent in 2001 and 2003. There was a drop in concern about security issues, from 37 percent in 2000 to 27 percent in 2003. In general, government administrators are less concerned about security and privacy than is the general public.[15]

In short, contrary to typical scholarly portraits of bureaucrats as hostile to change, the general disposition of administrators toward e-government is positive. Bureaucrats believe the Internet improves service delivery, performance, and efficiency. They have some fears about the security and privacy

TABLE 2-6
Top E-Government Obstacles, in View of Government Administrators, 2000–2003

	2000 (%)	2001 (%)	2003 (%)
Lack of Financial Resources	26	44	44
Security Issues	37	28	27
Entrenched Operating Procedures	13	19	19
E-Government Not a Priority	18	19	16
Inability to Recruit Qualified Personnel	23	26	10
Lack of Leadership Support	6	5	2

Source: Council for Excellence in Government National Surveys of Government Administrators

of digital technology, but they are nonetheless more confident than the general public about these problems.

CONSTRAINTS ON TECHNOLOGICAL CHANGE

From these bureaucrat surveys, the evidence is clear that government administrators are favorably predisposed to digital technology and e-government. Positive personal predilections toward technology are not, however, sufficient impetus for technological change. The capacity for change still is limited by factors such as the need for multiple service-delivery systems, bureaucratic fragmentation, budget resources, group conflict, media coverage, and partisan cleavages. Until these constraints are overcome, it will be difficult for government to realize the full potential of the Internet in service delivery and interactive democracy.

The Two-Systems Problem

In the early days of the Internet, politicians and proponents sold it as a revolutionary tool that would cut costs and improve the effectiveness of public sector performance.[16] In reality, though, neither has proven to be true. The up-front investment costs of new technology are substantial and cost savings do not emerge until enough users start taking advantage of electronic delivery systems that governments can save money through traditional bricks-and-mortar delivery systems. As long as government agencies have to maintain multiple delivery systems, service delivery is going to cost more because the infrastructure that provides online tax filing or electronic filing of citizen complaints exists on top of traditional government offices operating through face-to-face contact, phone calls, and mail.[17]

Not only do multiple channels increase costs, they create considerable potential for organizational conflict. People who work in digital government come from different educational backgrounds, organizational cultures, and salary situations than do the traditional staff responsible for processing phone and paper demands. Digital workers tend to be more highly educated, used to working in informal settings, and tend to be paid more than their traditional counterparts.

This clash of old- and new-economy perspectives complicates technology integration and makes it difficult for organizations to function effectively. Each side jockeys for political position and budgetary resources. These tensions slow the rate of technological innovation and prevent the public sector from taking full advantage of innovative features.

Bureaucratic Fragmentation

Another major issue in the application of new technology in the public sector is its adoption and diffusion among government agencies. As with many types of organization, public agencies sometimes are reluctant to change course and alter their ways of doing things. Despite the New Public Mangement, bureaucrats have set routines, and it is difficult to introduce new inventions that require people to be retrained and learn alternative techniques for fulfilling the agency's various missions. As pointed out earlier, technology-driven models resolve this "adoption" problem by merely assuming that the good qualities of technology will be readily apparent and government agencies will rush to embrace new creations. It is a deterministic perspective that equates invention with adoption, and ignores the organizational and political factors that either speed up or slow down the diffusion of innovation. Organizational factors are of particular importance when a new device comes along. It must be remembered that there is nothing immutable requiring bureaucrats to embrace it. Rather, officials must make concrete choices to share knowledge, adopt technology, and use technology to further their organization's goals.[18] The new technology must be integrated into organizational routines and improve citizen access to information and services.

Historically, the adoption process followed by organizations rarely has been simple or straightforward. Technology often has required a variety of accompanying actions that facilitate the adoption of new inventions. Even the most innovative creations need infrastructure, political will, and budget support in order to diffuse within the organization. Given the fact that e-government operates along different principles from the traditional government bureaucracy, the integration of technology into the public sector requires a novel way of thinking on the part of bureaucrats.[19] Rather than focusing on particular agencies and building their own infrastructure for information and service delivery, when it comes to e-government bureaucrats must work together across agency lines and build an infrastructure that is not particular to a single agency.[20]

In this situation, government officials interested in adopting technology must be forward-looking and collective-oriented. They must overcome the inertia typical in many offices and get people to think outside the box in terms of new ways of delivering services to the general public and the business community. It is not an easy task, which is one of the reasons why organizational and political constraints often slow down the adoption and diffusion of technology in the public sector.[21]

One of the goals of business models associated with New Public Management is to place external pressure on bureaucrats to improve governmental efficiency and effectiveness.[22] Through alterations of incentives and the creation of new mechanisms of organizational accountability, this approach seeks to change how bureaucrats think about their jobs and perform agency functions.

According to Louis Tornatzky and K. Klein, three qualities associated with public sector technological innovation are the ease of usage, compatibility with agency mission, and existence of a relative advantage in the use of the technology. Inventions that maximize these features are the ones most likely to be adopted by a government agency.[23] This finding is consistent with work by Daniel Bugler and Stuart Bretschneider as well as an empirical study by Charles Hinnant suggesting that organizations geared toward outside constituencies are the ones most likely to innovate.[24]

Jae Moon writes that there is a relationship between organizational size and type of government and the adoption of e-government. In his study of municipal governments across the country, he found that larger cities and those with professional management in the form of city managers were most likely to have a website. For example, 90 percent of council-manager governments had a site, compared to 77 percent of mayor-council governments.[25] This suggests that organizational qualities are important to technological innovation and that more professional bureaucracies are the ones most likely to innovate.

Budgetary Resources

A third factor that is crucial for e-government development is the cost of information technology.[26] New technology is very expensive, especially in terms of its up-front costs. A study of state web portal development by Diana Gant, Jon Gant, and Craig Johnson found that of the sixteen states about which they had data, the average cost of developing a state web portal was $2 million, with the range running from $303,250 to $6.5 million.[27] This demonstrates that e-government is quite expensive, and requires an infusion of substantial financial resources on the part of the government agency.

When in 2000 Pennsylvania developed its online system for renewing vehicle registration and driver's licenses, the contract to private vendor American Management System cost $3.5 million. During the first month of operation, 18,000 renewals were completed online. According to state officials, the cost for online registration is 50¢ per application, compared to $1.09 for mail renewal, and $7 for in-person registration at the state's Department of Transportation.[28]

Of course, development expenses for government websites eventually are recouped over time as costs get amortized and spread over more users. For example, the Virginia Department of Motor Vehicles estimates that personal visitors wishing to renew their car registration cost $5 to process, compared to $2.50 for those who register online. Initially, in May 1999, when online registration first was offered, 3,794 citizens used the service. This number rose to 21,000 people by December 2001, however, which helped to pay for the up-front technology investment.[29]

In the United States, state governments provide a relatively low portion (1 to 2 percent) of their overall budget to IT. Table 2-7 reports spending data provided by state chief information officers in twenty-three states where this information was available. The findings are consistent with budgeting literature that shows that most state budget categories were flat and incremental in terms of overall spending. Barring such examples as state corrections during the 1990s, which experienced rapid and substantial budget increases due to crackdowns on crime and the resulting increase in prison populations, most categories of government expenditures over the last decade remained fairly stable.[30]

Not only are IT expenditures comparatively small, but the numbers presented in this table demonstrate that spending was relatively uniform from 1998 to 2000. The bivariate Pearson correlations between IT expenditures in FY98 and FY99 is .92 (significant at .001), between FY99 and FY00 is .97 (significant at .001), and between FY98 and FY00 is .93 (significant at .001). This means at least 85 percent of the variation in FY99 spending can be predicted by what was spent in FY98. For FY00, one can predict 94 percent of the variation in state IT spending through FY99 expenditures. As suggested by Aaron Wildavksy's incremental budget model, the strongest predictor of one year's IT budget is that of the previous year.[31]

In looking at electronic governance, there are several models used to finance innovation in the e-government area: general tax revenues, user fees, and commercial advertising (or some combination of the various revenue sources).[32] The first model relies on general taxpayers to pay for the cost of new technology. In most states as well as in the federal government, this is the most common method for financing e-government. Electronic governance is seen as a collective benefit to the country, so it makes sense that its basic cost is paid for by taxpayers in general.

When economic times are good and governments have abundant resources, tax revenues are a popular way to pay for e-government. Budgetary trade-offs with other spending categories are less intense and groups outside of government can bargain and negotiate their share of government contracts. When times turn tough, however, spending on e-government must compete with expenditures for education, health care, and welfare. This is a much more difficult environment in which to rely on tax revenues, and as a result government officials typically demand greater evidence that spending in this area is worth the public investment.

A second model involves user fees, whereby those individuals accessing online information and services pay for the costs of providing those materials. The idea here is that a relatively small proportion of taxpayers access government services, and that at least in the short run, e-government will remain an elite activity benefitting a small number of users rather than an across-the-board advantage for everyone within society. Even with the most common type of online service, that of tax filing, only about a quarter to a third of the population uses online as opposed to paper filing.

The question in this situation is whether it is fair for the general taxpayer to

TABLE 2-7
Information Technology Expenditures as Percentage of Total State Budget,
1998–2000

State	FY98 (%)	FY99 (%)	FY00 (%)
Arizona	2.00	1.54	—
Colorado	2.66	2.24	2.52
Kansas	1.65	—	—
Kentucky	1.19	1.13	1.24
Louisiana	1.28	1.61	1.57
Maine	1.04	0.98	—
Maryland	2.14	2.74	2.88
Michigan	1.24	1.20	1.24
Mississippi	0.75	0.70	—
Nevada	1.55	1.72	1.66
New Jersey	1.56	1.96	1.93
North Carolina	—	0.66	—
North Dakota	3.62	3.30	3.40
Ohio	1.12	—	—
Pennsylvania	1.67	—	—
Rhode Island	0.80	0.79	0.69
South Dakota	—	2.76	3.26
Texas	2.75	3.06	2.89
Utah	1.93	1.87	1.52
Virginia	2.06	2.20	2.44
Washington	—	3.18	—
West Virginia	1.71	—	—
Wyoming	1.06	1.61	—
Mean	**1.69**	**1.85**	**2.09**
Standard Deviation	.72	.85	.86

Source: NASIRE 2000 Report on State Information Technology, Module One: Budgets, 2000

subsidize a service that is utilized by a relatively small group of citizens and a group that is relatively well educated and well off. Does it make more sense for the costs of e-government to be borne by those who actually are making use of it? The whole idea behind user fees is to link investment to those who are deriving benefits from specific services that are being used. During periods of government deficits, there is much more attention paid to user fees as an alternative way to finance e-government.

The third model, commercial advertising, extends the approach common with media outlets to the Internet. It suggests that e-government financing can be paid through the placement of product ads on government websites. Just as some city buses have ads on their sides, public sector sites can be financed through advertisements. The beauty of this approach is that it requires less public investment and relies instead on monies provided by the private sector; however, few city, state, or federal agencies rely on advertisements.

With government tax revenues weakening, some cities and states have turned to what is called a "self-funding" model partnership with the private sector. Under this arrangement, popularized by a company known as NIC, governments outsource portal development to this corporation in return for the company deriving income from user fees. This company first developed online applications in 1991 and now has contracts with seventeen states and eight local governments in the United States plus contracts with governments in Canada and the United Kingdom. Twelve states participate in the self-funding budget model that eliminates taxpayer financing of electronic government.[33]

As a financing mechanism, this partnership has the advantage of transferring the up-front development costs to the private sector; this means taxpayers do not have to pay for the construction of government portals. In return, however, NIC gets to keep user fees generated by website services; this means that the public sector loses revenue streams as e-government usage starts to take off.

According to corporate reports filed by the company, the quarter ending September 30, 2002, was the first time NIC generated a profit ($220,000) since it went public in 1999. Revenues increased by 15 percent to $8.9 million. The company said its gross margin for outsourced portals was 43 percent, up from 30 percent in the third quarter of 2001. Company chief executive officer Jeff Fraser noted that "the marketplace has become increasingly receptive to NIC's core self-funding business model because we help government leaders provide a new service delivery channel to citizens at no cost to taxpayers." He noted that NIC had developed 59 "revenue-generating applications" in the third quarter and had 80 more planned for the future. The reason why the company was able to increase its profitability was that it was able to deliver services through a fixed cost base. "NIC operates 26 separate portal installations, and the economies of scale are working in our favor," said chief financial officer Eric Bur. With government revenues falling short of expectations, this type of arrangement has become attractive to a variety of agencies.

Group Conflict and Outsourcing

A factor that constrains the degree of possible technical change is conflict between groups outside of government. In any political situation where governmental benefits are being allocated, private groups compete to win benefits.[34] It is an axiom of politics that interest groups want to win government decisions and contracts that are favorable to their interests. Groups fight over how much technical development work should be done in the private sector versus undertaken by the public sector itself.

In the United States, there is a bias in favor of relying upon the private marketplace as much as possible. Officials in public agencies often prefer to outsource tasks or rely on private vendors when they can. This holds down the size of the government bureaucracy and prevents the public sector from ballooning out of control. Only when the private sector is unable to complete a task or there are economies of scale or special needs do agencies undertake tasks themselves. The government can subsidize, form public/private partnerships, and regulate, but there always is a question of the extent to which it should take responsibility for technical innovation in the public sphere.

When Senator Joseph Lieberman introduced a bill, The E-Government Act, designed to make government purchasing of new computers more systematic and subject to uniform standards, industry groups indicated they thought the government should rely on the private sector for standardization. "The best way to go down that path is to let the commercial marketplace develop those tools. We don't need the federal government to develop interoperability," said Jeanne Foust, director of governmental affairs for software company ESRI.[35] This view was echoed by David LeDuc, director of public policy for the Software and Information Industry Association. He pointed out that "standards can inhibit growth in the industry. The standards the government sets are big enough to affect the industry."[36] Although the overall legislation passed, Lieberman's amendment on uniform standards was not included in the final bill.

Due to their expertise, financial clout, or political connections, some private groups are in a stronger position to compete for public benefits than others. This variation in competitiveness affects the extent to which technology gets incorporated in an agency's mission. The way governments structure these negotiations has tremendous consequences for various segments of the industry. Group conflict is particularly relevant for technological change because much of the innovation that takes place in any society comes from outside the public sector. Either private companies (sometimes with state support) or universities and nonprofit organizations are leading engines for creating new ideas. They are the part of society explicitly designed for knowledge creation and technology transfer to other organizations. These entities are the incubators that produce new inventions for government agencies.

Recognizing this fact, much of the public sector takes advantage of research assistance from outside of government in order to boost productivity and deliver services more effectively. It is hard for government bureaucrats to innovate on their own. A more common model is for private or nonprofit groups from outside the public sector to lobby for change and provide expertise that helps bureaucrats adopt new technology in particular agencies.

In the e-government area, there are several different ways in which new technologies have been introduced into the public sector: in-house development by the government agency, outsourcing to the private sector, and public/private partnerships with nonprofit groups. According to the study by Gant, Gant, and Johnson, 43 percent of American states relied on private firms to develop their portal, 33 percent did it in-house, and 24 percent used some type of public/private partnership.[37] This demonstrates that, overall, private groups play a role in two-thirds of new government portals.

This is consistent with survey work by Peter Hart and Bob Teeter. In their 2003 study of government administrators, they asked how much of the development or maintenance of their agency's e-government applications has been contracted to private companies. Only 27 percent reported that none of the applications were outsourced. Of those that were contracted out, 3 percent of administrators said all of their applications were done by private contractors, 10 percent said most of it was, 36 percent stated that some of it was, and 14 percent reported that only a little of it was outsourced.[38]

The particular way in which groups participate in website development is important because each of the approaches noted above involves varying degrees of reliance on group expertise and access. In-house development, for example, creates the least potential for outside group influence because the public sector handles implementation itself without relying much on formal help from external organizations. Typically, in this situation, government officials rely on IT experts already hired by the public sector to develop e-government systems. These individuals develop and maintain government websites and deal with troublesome implementation issues as they arise.

If the system being developed is relatively simple, in-house development is a perfectly good way to implement e-government; Public officials merely work with those they already know. There is little need for outside expertise, costs are covered by line items that already appear within the agency budget, and the agency retains control over implementation of the technology.

If the system is complex, however, governments generally have to rely on expertise from outside of the public sector. This involves some loss of control and decline in autonomy on the part of the bureaucratic agency. Private actors are more important in this kind of innovation, and government authority is limited to more of a coordination or oversight role.

In situations where the agency is developing e-government portals or adding complex services online, outsourcing is a more common model of technological change.[39] The public sector agency puts out bids directly to private groups for the right to develop the digital infrastructure. Through this competitive bidding process, groups compete for government contracts that allow them to build a portal and/or add services to an existing portal.

This model is popular because many government agencies lack the expertise or time to develop the hardware and software necessary for e-government.[40] Given the lack of available staff, they turn to private companies with expertise to build new software systems. Subject to agency oversight, these groups develop and implement the electronic service delivery that then becomes available to website visitors.

Most government agencies rely on outsourcing for portal development and to add services to existing government websites. These tasks tend to be rather technical so officials prefer to rely on the private sector for these enhancements rather than using staff already on hand within government. As will be discussed in chapter 5, contracts vary depending on the type of service being added. Outside groups typically are hired to construct sites, add services, or make an existing site more user-friendly and accessible to the disabled.

Public/private partnerships are still another way e-government is implemented.[41] Sometimes, the public sector works coooperatively with universities or nonprofit groups to develop government websites. These groups provide the expertise necessary for the incorporation of the new technology into the public sector. It is a way to expand the range of expertise available within government offices and give officials access to the best available talent.

FirstGov.gov, the original American national government website, is the most famous example of this approach. That site was first developed in 2000 by a private individual, Eric Brewer, who worked at Inktomi Corp., a website development company. Brewer had mentioned the idea of a national government portal to President Bill Clinton during a conversation lasting a "minute or so" in 1999 at the World Economic Forum in Switzerland.[42] Working in conjunction with a $4.1 million federal grant to GRC International of Vienna and the nonprofit Federal Search Foundation that put together the site's search engine "FedSearch," Brewer built the site and then turned it over to the General Services Administration (GSA) for updating and implementation. The GSA now operates the site on behalf of the national government.

This approach allowed the government to get a site up and running quickly without having to rely on government employees or go through a lengthy bidding process. In the space of a few months, Brewer developed the site and built links to "20,000 U.S. government Internet sites and 27 million pages of data."[43] It proved to be one of the most frequently visited government websites in the country, and serves as the major gateway for the United States government.

Two years later, Vice President Dick Cheney officially launched a new version of the site saying the goal of the portal was to "remedy one of government's oldest problems: the slow, confusing and often ineffective ways in which it responds to the public." Costing $350,000 and overseen by the GSA, FirstGov boasted as its motto, "Three Clicks to Service."[44] With a new design and more online information and services, FirstGov has become very popular.

Each of the three development models (in-house, oursourcing, and public-private/partnerships) varies considerably in the areas of reliance on outside groups and the degree of conflict with outside sources. In-house models generally involve the least group conflict because government officials are developing systems using existing personnel. Outsourcing engenders more conflict because groups compete for the right to build government websites. Public/private partnerships involve some degree of conflict but this is mediated by the fact that many of the alliances come with nonprofit rather than for-profit organizations. The nonprofit nature of this group competition reduces the most vociferous type of conflict and thereby facilitates the speedy adoption of new technology by the public sector.

Leadership, Partisan Conflict, and Media Coverage

The last set of factors that affects the adoption of new technology is the degree of leadership, partisan conflict, and media coverage. Political leadership and professional staff are vital for bureaucratic inertia to be overcome. Leaders use the media to highlight top priorities and marshal the financial resources that are necessary for the implementation of technical improvements. Those who are interested in making a big leap forward in terms of technology typically need to convince other people that it is more important to spend money on technology than other important priorities, such as health, education, or welfare.

Indeed, the professionalism of the state has been found to be related to the adoption of e-government services. A study by Ramona McNeal, Caroline Tolbert, Karen Mossberger, and Lisa Dotterweich found that "innovation in e-government is driven by legislative professionalism and to a lesser extent, state professional networks."[45] States that ranked highly on legislative staff, salary, and length of legislative session were more likely than others to have online services. Research by Paul Ferber, Frank Foltz, and Rudy Pugliese on state legislative websites found a moderate correlation between legislative staff per capita and quality of websites maintained by legislatures.[46]

Technology obviously is not a high priority for all officials. Some see it as an expensive and untested investment that could yield few dividends. Others merely see alternative issues as being more important. Still others bow to the opinions of vested interests, who fear that new technology will alter the status quo and thereby damage their own prospects. In cases where governments make progress on technical innovation, however, there generally is a governor, mayor, or legislator committed to action in this area. While it is hard to measure an intangible factor such as leadership, case studies demonstrate the importance of individual initiative for the marshalling of financial resources and getting government agencies to work together.[47]

For example, Wisconsin and Arkansas represent examples of states that have devoted extensive resources to e-government development in recent years.[48] In both cases, there were governors (Scott McCallum in Wisconsin and Mike Huckabee in Arkansas) who believed it was important to innovate technologically. Each of these elected officials made speeches about the importance of electronic governance, lobbied the legislature, and conducted interviews with media reporters emphasizing the importance of this new priority. Due in part to these efforts, both men were able to overcome bureaucratic intransigence and make progress on e-government. In McCallum's case, he even was able to get his state's legislature to create a new Department of Electronic Government, the first state in the nation to do this.[49]

Conversely, in places where there has been slow progress, political leaders generally have not made e-government a very high priority or devoted much in the way of personal political capital to improving online service delivery. For example, in 2001, when the city government website for Albany, New York, was found to rank 70th out of the 70 largest metropolitan areas in the country, its mayor joked to a newspaper reporter that he didn't even use email and that he preferred "to do business the old-fashioned way." Indeed, he noted, "I prefer the personal touch."[50] The same reaction was evident in other places where governments ranked poorly on electronic governance. Many political leaders did not consider the Internet to be a very important priority or a device for revitalizing the public sector.

In the United States, e-government generally has been presented as a technocratic "good government" reform. Unlike education, housing, and health care, the technology area has not been buffeted by many partisan controversies or been politicized by opposition leaders. In the early days, for example, there was a professional consensus among government technicians that electronic governance was not a liberal or conservative issue or a Republican or Democratic subject. Rather, it was a neutral technology that offered the potential for improving efficiency and effectiveness in the public sector.[51]

This "technocratic" vision helped build support for e-government across party lines and allowed policy entrepreneurs to marshal resources necessary for its implementation.[52] Technology proponents saw electronic government as a way to overcome past ideological controversies about the proper role of government. Such a perspective fit with the classic view of technology as an enabler of scientific progress and not something that would divide people. After all, when the automobile was introduced, it did not become a political issue between the parties. Instead, this new invention was touted as something that would ease the lives of all who could afford it. Eventually, as cars became more prevalent, they were identified as an incarnation of the American Dream. The same was true for television and telephones. None of these creations became political footballs, but were sold as nonpartisan consumer devices that would improve the lives of everyone.

In recent years, however, there is evidence of partisan controversy and media scandal with respect to e-government. These conflicts highlight the role of politicians and the media in shaping the political context for e-government. In 2002, the technocratic consensus over e-government was undermined when funding and implementation of Internet technology became partisan issues in gubernatorial races around the country. Following the public release of a Brown University study on state and federal e-government, Governor Mike Huckabee touted Arkansas' progress in this area. Speaking to his department heads, Governor Huckabee noted that Arkansas ranked number one in providing electronic services (180 online services in all, from tax filing and hunting licenses to car registration and corporate filngs). "This is outstanding news," he announced. "We've been touting for quite some time the importance of e-government—moving Arkansas to a place where we're better serving our citizens by going where they are rather than making them come to where we are."[53]

Shortly thereafter, his Democratic opponent Jimmie Lou Fisher complained that Governor Huckabee had not told the entire story. "What he's not telling you is that in the same study, Arkansas dropped from 19th to 30th in overall e-government ratings," said Fisher campaign manager Vincent Insalaco. "We are assuming that in part it's because of the failings of the AASIS computer system." According to newspaper reports, the Arkansas Administrative Statewide Information Systems (AASIS) had encountered serious problems with vendor payments and security over the course of the preceding year.[54] Despite the controversy, Governor Huckabee was re-elected by a 53 to 47 percent margin.

Arkansas was not alone in seeing gubernatorial controversies about e-government. In Wisconsin, Democratic challenger Jim Doyle blasted Governor Scott McCallum after that state fell from 26 in 2001 to 46 in 2002 (of the 50 states). Referring to the new cabinet office created to administer e-government, Doyle campaign director Bill Christofferson said, "The Department of Electronic Government is a symbol of McCallum's failed policies." It was a "colossal mistake," the campaign manager noted and argued that the department had been criticized for $11.4 million in overspending on outside consultants. If elected governor, Doyle said he would eliminate the department.[55]

McCallum's spokesperson Debbie Monterrey-Millett defended the governor's stewardship of e-government. "If Jim Doyle deleted the Department of Electronic Government, all state agencies and departments would have to go back to the old way of doing things," she said. The department has saved more than $18 million for taxpayers, she indicated. She also complained about the loss of the bipartisan spirit that had marked the early days of the department's creation. "It wasn't until the silly season of the campaign began that some Democrats decided to attack it," she said.[56] In the end, challenger Doyle unseated incumbent governor McCallum by 45 to 41 percent.

With e-government emerging as a partisan issue between some Democrats and Republicans, it is not clear whether digital democracy can retain its veneer of nonpolitical, technocratic government reform. When politicians start arguing about overspending by outside consultants and the way e-government is being implemented, and reporters write critical stories about contract procurement, it becomes much more difficult for the public to see the subject as a nonpartisan issue. Rather than being a device for revitalizing citizen trust in government, e-government becomes just another sign of inefficiency in the public sector. To the extent that voters associate it with cost overruns and partisan controversies, e-government will lose its luster as a technocratic solution for improving government performance.

The information on government websites also is becoming an object of controversy between outside interest groups. Issues such as abortion long have generated extensive diviseness between those interested in a woman's right to choose and disapproval of the termination of unborn fetuses. Ever since the landmark U.S. Supreme Court decision in *Roe v. Wade* in 1973, each side has fought a series of large and small battles over their respective perspective on this social topic.

It therefore is no surprise that as governments have added information to public sector websites the abortion controversy has become entangled in the development of e-government. For example, the National Cancer Institute website used to note that scientific research demonstrated "no association between abortion and breast cancer," but under the Bush Administration changed the language to say the current evidence was "inconclusive" following complaints by conservative members of Congress. In addition, a webpage for the Centers for Disease Control and Prevention previously wrote that "studies showed that education about condom use did not lead to earlier or increased sexual activity." Responding to complaints from "abstinence only" groups, however, that conclusion was deleted during the George W. Bush Administration.[57]

These removals of online information from government sites ignited a firestorm of political criticism about control of public sector websites. According to critics, the administration "is now censoring the scientific information about condoms it makes available to the public." Rather than information be openly available, House Democrats complained, "information that used to be based on science is being systematically removed from the public when it conflicts with the administration's political agenda."[58]

This battle is emblematic of larger conflicts over the dissemination of government information via the Web. With newfound concern over homeland security, military sites have removed material on troop movements and public sector agencies have eliminated sensitive information about the location of power plants and other sites potentially of interest to terrorists. For example, the Environmental Protection Agency no longer lists which American manufacturing plants store chemicals. The Nuclear Regulatory Commission deleted its Nuclear Materials Events Database that provides an inventory of radioactive materials in the United States.[59] These moves were in keeping with more general concern about terrorism and security following the September 11, 2001, terrorist attacks in New York City and Washington, D.C.

In the early days of e-government, there was a headlong push to put information online. The idea was that electronic governance was a technocratic reform with no political or ideological agenda other than making government operate more efficiently and effectively. Due to the principle of the public's right to know, government officials worked hard to put more and more information online. Now, due to partisan battles concerning the threat of terrorism, the placement or removal of information on government websites is seen as a more political activity and is subject to group battles and the differing ideological stances of various officials. Much like the rest of government, the technocratic vision is being replaced by the political fight over ideas. Groups are contesting website statements and databases in the same way that previously they would argue over a written report or committee recommendation.

While this represents a natural progression of new technology, it risks the erosion of public confidence in e-government. As long as the public sees e-government as nonideological, large segments of the population will view it in nonpolitical terms. They will support new spending on technology and encourage administrators to move forward in their attempts to bring more online services to websites.

As the public comes to believe that e-government is subject to the same political forces as traditional government, however, there is the danger that citizens will grow cynical about electronic democracy and no longer will view it as a technocratic mechanism for the improvement of government service delivery. Contracting scandals and charges of favoritism in vendor awards attract critical scrutiny from journalists and lead to stories that undermine the technocratic image of government reform. In the long run, this weakens support for e-government and makes it more difficult to attract financial resources to this area.

The Content of American Government Websites

ONE OF THE CHALLENGES in assessing e-government is the absence of an agreed-upon consensus as to what constitutes successful performance. Befitting a field that still is in its infancy, there are many different methodologies and standards for measuring online government. Some rely on technical performance, such as download times. An innovative project by Irina Ceaparu and Ben Shneiderman, for example, compared the downloading of government websites from the fifty American states and discovered that Vermont downloaded the smallest total byte count (42K), while Washington had the largest (274K).[1]

Others have employed public opinion surveys to document usage. One illustration is a project undertaken by a consulting firm, Taylor Nelson Sofres. Its researchers interviewed over 29,000 adults in twenty-eight countries on two different questions: how many people use e-government and what fears users have about providing online personal information to the government. Based on these responses, the analysts rated the countries in terms of online government usage. The countries with the highest percentage of users were Sweden, Norway, Singapore, Denmark, Faroe Islands, Finland, Canada, Australia, and the United States.[2]

Still other approaches rely on the perceptions of government officials about their websites. This is a technique employed by the annual Digital State Survey (2000–2003) undertaken by the Center for Digital Government and the Progress and Freedom Foundation.[3] In this study, which is sponsored by technology firms such as Microsoft, Hewlett-Packard, and AMS, researchers interview each governor and state chief information officer about their website's contents. Among other things, this project quizzes officials about eight different kinds of innovation, such as education, geographic information systems, digital democracy, taxation and revenue, electronic commerce, and social services. States are given up to 100 points in each area, and ranked for the degree of technological progress that is reported.

Based on this research, the foundation claims that the top e-government innovators are Arizona, Michigan, Washington, Illinois, Wisconsin, Virginia, Utah, Indiana, South Dakota, with Connecticut and Maryland tied for tenth. In its press release distributed with the 2002 report, the Center for Digital Government noted that states with political leadership highly committed to technological change were the ones that did well. Cathilea Robinnett, executive director of the Center, explained Arizona's top-place finish as indicative of its "long-term commitment to electronic service delivery." In Michigan, the second-place finisher, the release cited the crucial role of Governor John Engler in technology, who "personally drove much of Michigan's e-government progress."[4]

While useful in some respects, the obvious limitation of a survey approach is its reliance on perceptions and self-reports, as opposed to actual online content. Government officials have incentives to report results in the most favorable light in order to enhance their state's reputation. This skews indicators in a positive direction and underplays constraints on technological change. In addition, not all states participate in the study. In 2002, for example, five of the fifty states declined interviews, and therefore were not ranked.

Another approach looks at actual website content and assesses e-government based on what is online. A content-based perspective approximates the experience of citizens and businesses that go to websites. Visitors are not interested in how officials justify e-government, but in what is actually accessible. By looking at specific content, it is possible to examine what kinds of information and services are online as well as how issues of privacy and security are handled.

Of course, focusing on content does not solve the problem of which indicators to use. In looking at projects that examine website content, there are substantial differences in the extent to which the assessment emphasizes services, interactivity, accessibility, or democratic capacity. In the language of the four-stage e-government model, some observers focus on stage three (service delivery and portal integration) while others employ the standard of stage four (interactive democracy and citizen accessibility). Both dimensions, of course, are important to the evaluation of digital government.

An annual study undertaken by the consulting firm Accenture from 2000 to 2003 of twenty-two nations around the world emphasizes a service model centering on "customer relationship management." This is a philosophy based on approaching e-government's success in meeting citizen needs and maximizing government efficiency. According to Accenture, the services that should be placed online are those "that provide a real return on investment, either through increased service effectiveness or efficiency."[5]

Countries were rated based on how "successfully they adapted aggressive eGovernment services for their citizens." In the 2002 study, Canada was rated number one, followed by Singapore, the United States, Denmark, Australia, Finland, Hong Kong, the United Kingdom, Belgium, Germany, Ireland, and France. Consistent with the need of a private company seeking to expand its market, the Accenture study is service-oriented. It emphasizes having "a broad, cross-agency approach to providing government services through multiple delivery channels. The government portal provides information and services not only by subject and audience, but by department and agency as well."[6]

The Accenture project does not measure the extent to which means of public feedback are incorporated onto websites, how accessible they are, and whether they include opportunities for public deliberation or information that would help citizens hold leaders accountable. In the language of the four-stage e-government model, the Accenture research emphasizes stage three, fully executable portals, not the fourth stage, interactive democracy.

In contrast, the Cyberspace Policy Research Group headed by Chris Demchak, Christian Friis, and Todd LaPorte focuses on the organizational qualities of government websites.[7] With support from the National Science Foundation, this group studied over forty indicators of website openness, transparency, interactivity, and effectiveness. Among specific items examined were contact information, issue information about the agency, security statements, privacy statements, agency mission statements, reports, searchable indices, electronic updates, foreign language access, and disability access. Based on these indicators, the group concluded that the United States, Canada, Israel, Taiwan, India, Australia, United Kingdom, and Malaysia comprised the top performers in terms of number of government websites.

A 2001 analysis undertaken by the United Nations Division for Public Economics and Public Administration and the American Society for Public Administration represents still another way of looking at website content. This project looked at the 190 United Nations member countries, and focused on the degree to which citizens could conduct transactions online in five critical sectors: education, health, labor/employment, welfare/social services, and financial services. Among the top countries identified were the United States, Australia, New Zealand, Singapore, Norway, Canada, United Kingdom, Netherlands, Denmark, and Germany.[8] Rather than looking across the board, it focused on a smaller set of e-government areas. In assessing these websites, it examined availability of online services, access to databases, foreign language translation, search capabilities, areas to post comments, chat rooms, digital signatures, and media streaming. It did not, however, look at accessibility features for the disabled or readability levels. Neither did it explore interactive features, such as website personalization and electronic updates, that are central to Internet technology.

Other research projects look mainly at the United States and focus on particular agencies or individual states. For example, Paul Ferber, Franz Foltz, and Rudy Pugliese, Jody and Bryan Fagan, and Matt Carter and Ryan Turner focus on state legislative sites to see how much information is contained there.[9] The Congressional Management Foundation and the Graduate School of Political Management at George Washington University used a team to study five building blocks (audience, content, interactivity, usability, and innovativeness) of congressional websites.[10] Juliet Musso, Christopher Weare, and Matt Hale assessed California websites for the degree to which they focus on management and services versus democracy.[11] The Wisconsin State Legislative Audit Committee studied local government in its state to see what services were online.[12]

In looking at these various studies, there is little unanimity on what areas of e-government to emphasize.[13] Some look across the board at a wide range of agencies. Others focus on selected subsets. Some emphasize service or usability, while others prefer to investigate organizational features or perceptions about electronic government. Not surprisingly in this situation, there is little consistency in the states, agencies, or countries that receive top marks.

Later in this chapter, I adopt a content-based approach and focus on the extent to which the public sector has succeeded in putting information, services, and databases online, as well as incorporating interactive features designed to improve accountability and democratic outreach. These features are fundamental to public sector performance and central to theories how citizens link with leaders in a democracy. Scholars cannot answer important questions about e-government without studying website content and how things have shifted over time.

To study digital government, I examine the content of 11,426 American government websites at the city, state, and federal levels in 2000, 2001, 2002, and 2003 (see chapter 9 for a discussion of the content of foreign national government websites). As explained in chapter 1, in 2000, I looked at 1,716 state websites and 97 federal websites, while in 2001, I investigated 1,621 state websites, 58 federal websites, and 1,506 city websites. In 2002, I analyzed 1,206 state government websites, 59 federal websites, and 1,567 city websites. In 2003, I investigated 1,603 state websites, 60 federal websites, and 1,933 city websites.[14]

I use this content analysis to investigate broader models of technological change. I define models of incremental, secular, or transformational change in two ways. I use the four stages of e-government (billboards, partial service delivery, portals with fully executable services, and interactive democracy) to distinguish the scope of change. As discussed in the opening chapter, incremental change includes public sector websites that have not moved beyond the partial service-delivery model. They either are stuck in the billboard stage and have online publications, databases, and agency contact information or they have progressed no further than partial service delivery. They have few services or interactive features, and do not pay close attention to privacy or security. Secular change is represented by websites that have moved into portals with a number of fully executable and integrated services. These sites are well integrated and easy to navigate. They take privacy and security concerns seriously. Transformational change is measured by sites that meet the requirements of interactive democracy. They have lots of online services, interactive features that enhance demoracy, website personalization, and avenues for public feedback and deliberation.[15]

In addition, I gauge the degree of change by looking at shifts for specific indicators from 2000 to 2003. As discussed in appendix I, my content review focused on more than two dozen different aspects of e-government performance from information availability and service delivery to privacy, security, disability access, readability, and foreign language translation, among others. These measures provide a means for looking at the pace and breadth of e-government performance in selected areas as well as the speed of overall digital improvement.[16]

The research results show steady improvement between 2000 and 2003 in online information (especially publications and databases), but no major transformation in how government provides services and conducts its business. Few services are online in most agencies and many governmental units have not incorporated features of interactive democracy that would improve the overall functioning of the political system. Of the three levels of government, the national government has made the greatest progress, followed by states and cities. There is considerable variation within each level of government, however, as well as across agency types. The progress that is present is consistent with a "checkerboard" perspective, with lots of divergence in technology innovation. This suggests an interpretation more in line with incrementalism and secular change than transformation.

ONLINE INFORMATION

The first subject examined is the availability of basic information at American government websites, and how patterns have shifted between 2000 and 2003. Befitting the requirements of a billboard stage, contact information and access to publications and databases are quite prevalent (table 3-1). The vast majority of state and federal sites provide their department's telephone number (96 percent in 2002) and address (95 percent). There also is abundant evidence of publications (98 percent in 2003) and databases (80 percent). About three-quarters of websites have links outside their agency. Similar to the pattern in previous years, most websites do not incorporate audio clips or video clips into their sites. In 2003, only eight percent provided audio clips and 10 percent have video clips.

In looking at trends over time, there is little evidence of radical change from year to year. Most sites incorporate contact and content in an incremental fashion, with few big changes over the four-year period. One of the bigger changes came in regard to databases, which doubled its count from 2000 to 2003. However, the lack of widespread change on many of these measures is consistent with an incrementalist view of technological change.

SERVICES PROVIDED

Placing services online is a top priority for many government planners. As websites evolve from the billboard stage to partial service delivery and fully

TABLE 3-1
Percentage of State and Federal Websites Offering Publications and Databases, 2000–2003

	2000 (%)	2001 (%)	2002 (%)	2003 (%)
Telephone Contact Information	91	94	96	—
Address Information	88	93	95	—
Links to Other Sites	80	69	71	—
Publications	74	93	93	98
Databases	42	54	57	80
Index	33	99	—	—
Audio Clips	5	6	6	8
Video Clips	4	9	8	10

Source: Author's e-government content analysis database

functioning portals, services become more paramount. Since citizens are used to being able to order books and other merchandise online at commercial sites, they want the same convenience from the public sector. Purchasing hunting and fishing licenses is very popular on government websites as is dealing with routinized government features such as paying fines for overdue parking tickets.[17]

Fully executable, online service delivery benefits both government and its constituents. In the long run, such services offer the potential to lower the costs of service delivery and to make services more widely accessible to the general public, who no longer have to visit, write, or call an agency in order to execute a specific service. Citizens enjoy the convenience of online services and governments use technology to deal with simple transactions that occur time and time again. Businesspeople appreciate being able to order permits online and request information that is important to their performance.

As more and more services are put online, technology proponents claim that e-government will revolutionize the relationship between government and citizens.[18] By improving service delivery at lower costs, these individuals suggest technology can improve citizen access and in the long run close the digital divide between information users and nonusers. It is, however, an empirical question whether government sites actually are offering online services.

Of the state and federal websites examined in 2003, 44 percent offer services that are fully executable online (table 3-2). This is nearly double the 23 percent that had online services in 2002, 25 percent in 2001, and 21 percent in 2000. Of the 2003 sites, 56 percent have no services, 15 percent offer one service, 8 percent have two services, and 21 percent have three or more services. Clearly, government websites are placing a higher priority in getting services online. Almost half of the websites do not have *any* electronic ser-

TABLE 3-2
Percentage of State and Federal Websites Offering Online Services, 2000–2003

	2000 (%)	2001 (%)	2002 (%)	2003 (%)
No Services	78	75	77	56
One Service	16	15	12	15
Two Services	3	4	4	8
Three or More Services	2	6	7	21

Source: Author's e-government content analysis database

vices, however. Reflective of the obstacles that limit technological change, relatively few agencies offer more than a handful of specific services.

Common services that are available online include: ordering a copy of birth or death certificates; filing consumer complaints; filing business and payroll taxes; placing new hire reports; updating professional licenses; filing Uniform Commercial Code reports online; reserving a hotel or campsite; purchasing hunting, fishing, or sporting licenses; renewing motorboat/snow-mobile/all-terrain vehicle registrations; renewing drivers' licenses; paying speeding tickets; renewing car registrations; ordering duplicate drivers' licenses; ordering special plates; purchasing transportation passes; ordering duplicate registration for motor vehicles; registering to vote, and subscribing to national "Do Not Call" registries.

Some states offer a number of innovative features. For example, Maine and Virginia have a "live" help feature in which citizens can get instant help from a real person if they encounter a problem at that website. The Washington portal has six foreign language options, including Cambodian, Vietnamese, and Mandarin Chinese. Florida offers documents in English, Spanish, and Creole. The Alaska Department of Motor Vehicles waiting room has an on-line webcam so people can see in real time how crowded lines are and thereby judge when a good time to go would be. The Virginia portal has wireless access on its portal site.

The moderate pace of expansion of e-government services demonstrates the obstacles that limit dramatic change through technology. As noted in chapter 2, there are organizational and political factors that constrain techno-logical change in the public sector. It is difficult for bureaucrats to alter routines and incorporate new technology in the agency mission. In addition, interest group conflict and lack of political leadership impede the diffusion of new technology. Without strong leadership, the fragmentation of bureaucratic life limits the diffusion of innovation. At the same time, the high costs of new technology restrain public sector innovation. It is expensive to place services online and as agencies create digital services, there is an even greater need for secure and confidential transactions involving social security and credit card numbers. Given these difficulties, it is not surprising that there has been limited change in the introduction of online services.

Another factor that has limited the expansion of online services is the lack of a means to pay for services through credit cards. It is common practice for commercial sites to offer goods and services online through the use of credit cards. Of the government websites analyzed, however, only 19 percent accepted credit cards in 2003. This number is double the 10 percent found in 2002 and 2001, and larger than the 3 percent in 2000. This increase in the number of sites allowing for credit card payments shows that online financial transactions are becoming more accepted. But the vast preponderance of government websites do not offer this payment mechanism, which limits the growth capacity of online services.

Despite federal legislation authorizing digital signatures for financial transactions, less than 1 percent have incorporated this technology into their sites. This is about the same as in the preceding three years. The inability to allow digital signatures limits the capacity of government websites to expand services and other kinds of legal transactions because many of these activities require a formal signature in order to execute the action. Until this capability is developed, citizens are limited to printing forms from governmental websites and mailing them into government centers with a handwritten signature.

Of the fifty states and the federal government, there is wide variance in the percentage of websites offering online services. Table 3-3 shows the average number of online services found per site in various states and in the federal government. I tabulated the number of services at each state agency and averaged the total over the number of agencies. Based on this analysis, Massachusetts is the clear leader, with an average of 25.4 online services across its websites. This is followed by Kansas (5.4 services), the United States national government (4.8 services), New York (4.7 services), Maine (4.2 services), and Louisiana (3.8 services). The states having the lowest average number of online services are Alaska (0.2 services), Wyoming (0.3 services), and New Mexico (0.5 services).

This variation in state performance with respect to e-government demonstrates how organizational and institutional factors in the public sector limit the diffusion of technology. Models based on technological determinism assume that whatever merits accompany a particular invention will be apparent to all and good ideas therefore will quickly be introduced throughout the land. That model clearly ignores how a range of factors within individual states, such as fiscal resources, political leadership, institutional conflict, and group competition inhibits the spread of technology across governmental units. Even several years after the incorporation of the Internet as a service-delivery mechanism, states vary widely in their ability to use technology to deliver services to citizens and businesses. In the next chapter, I analyze models

Table 3-3
Average Number of Online Services at State and Federal Government
Websites, 2003

Massachusetts	25.4	South Dakota	1.7
Kansas	5.4	Arizona	1.6
New York	4.7	Georgia	1.6
Maine	4.2	Ohio	1.5
Louisiana	3.8	North Carolina	1.5
Minnesota	3.8	South Carolina	1.4
California	3.4	West Virginia	1.1
Idaho	3.3	Colorado	1.1
Kentucky	3.3	Oregon	1.1
Florida	3.3	Hawaii	1.0
Maryland	3.2	Nebraska	1.0
Michigan	3.1	New Hampshire	1.0
Iowa	3.0	North Dakota	0.9
Missouri	2.8	Oklahoma	0.9
Texas	2.7	Nevada	0.8
Indiana	2.7	Montana	0.8
Delaware	2.5	Vermont	0.8
Pennsylvania	2.5	Wisconsin	0.7
Virginia	2.5	Rhode Island	0.7
Washington	2.3	Utah	0.7
Arkansas	2.2	Alabama	0.6
New Jersey	2.1	New Mexico	0.5
Mississippi	2.0	Wyoming	0.3
Connecticut	2.0	Alaska	0.2
Illinois	1.8		
Tennessee	1.8	**United States**	**4.8**

Source: Author's e-government content analysis database

that explain variation in the number of online services offered across the fifty states.

PRIVACY AND SECURITY

In addition to services and information availability, there are other features of e-government that are important to ordinary citizens. The virtually unregulated atmosphere of the Internet has prompted many to question the privacy and security of government websites. Public opinion surveys place these worries near the top of the list of citizen concerns about e-government. Having visible statements outlining what the site is doing is a valuable asset for reassuring a fearful population.

TABLE 3-4
Percentage of State and Federal Government Websites with Privacy and Security
Policies, 2000–2003

	2000 (%)	2001 (%)	2002 (%)	2003 (%)
Privacy Policies	7	28	43	54
Security Policies	5	18	34	37

Source: Author's e-government content analysis database

A growing number of state and federal sites offer privacy and security statements, as demonstrated in table 3-4. In 2003, 54 percent have some form of
privacy policy on their site, up from 43 percent in 2002, 28 percent in 2001,
and 7 percent in 2000. Thirty-seven percent now have a visible security policy, up from 34 percent in 2002, 18 percent in 2001, and 5 percent in 2000.
The increase in the percentage of sites with privacy and security policies is
due not just to the worries of citizens and bureaucrats about these topics, but
also the galvanizing effect of the September 11, 2001, terrorist attacks on the
United States. The airliner crashes in New York City and Washington, D.C.,
demonstrated major lapses in American security and persuaded government
officials to place a much higher priority on electronic privacy and security.
The speed with which progress has been made in this area suggests that
where there is a political will and where financial resources are freed up for
specific improvements, such as homeland security, technological innovation
can diffuse quickly. The exceptional nature of these improvements, however,
shows that it takes unusual circumstances (that is, major terrorist attacks) for
the government to move quickly to improve security.

In order to assess the quality of privacy and security, we evaluate the content of these publicly posted statements. For privacy policies, we look at several features: whether the privacy statement prohibits commercial marketing of visitor information; use of cookies or individual profiles of visitors;
disclosure of personal information without the prior consent of the visitor,
or disclosure of visitor information to law enforcement agents. As shown in
table 3-5, there has been a decrease in the degree to which consumer interests are protected compared to previous years. For example, whereas 39 percent of government websites in 2002 prohibit the commercial marketing of
visitor information, this year that number dropped to 32 percent. The same
is true for policies that prohibit the disclosure of personal information. In
2002, 36 percent of sites have this feature, but that figure declined to 31
percent this year. The relatively low numbers in these areas show that public sector sites have a long way to go to address specific citizen concerns
about digital government.

Table 3-5
Assessment of E-Government Privacy and Security Statements, 2001–2003

	2001 (%)	2002 (%)	2003 (%)
Prohibit Commercial Marketing	12	39	32
Prohibit Cookies	10	6	10
Prohibit Sharing Personal Information	13	36	31
Share Information with Law Enforcement	—	35	35
Use Computer Software to Monitor Traffic	8	37	24

Source: Author's e-government content analysis database

READABILITY

Literacy is the ability to read and understand written information. According to national statistics, about half of the American population reads at the eighth-grade level or lower.[19] A number of writers have evaluated text from health warning labels to government documents to see if they are written at a level that can be understood by citizens. The fear, of course, is that too many government documents and information sources are written at too high of a level for many citizens to comprehend.

To see how government websites fared, we undertook a test of the grade-level readability of the front page of each state and federal government website that we studied. Our procedure was to use the Flesch-Kincaid standard to judge each site's readability level. The Flesch-Kincaid test is a standard reading-tool evaluator used by the United States Department of Defense. It is computed by dividing the average sentence length (number of words divided by number of sentences) by the average number of syllables per word (number of syllables divided by the number of words).

As shown in table 3-6, the average-grade readability level of American state and federal websites was at the eleventh grade, which is well above the comprehension of the typical American. Sixty-seven percent of sites read at the twelfth-grade level. Only 12 percent fell at the eighth-grade level or below, which is the reading level of half the American public.

There were some differences between state and federal sites. Sixty-eight percent of state sites read at the twelfth-grade level, while 63 percent of the federal sites did so. It mattered a bit what the branch of government was. Sixty-nine percent of executive-branch sites were written at the twelfth-grade level, compared to 65 percent of legislative sites, 60 percent of judicial sites, and 56 percent of portal sites.

Agency type mattered much more, although not always in a manner consistent with the particular audience served by the website. One might expect

TABLE 3-6
Percentage of Government Websites Falling within Each Grade Level, 2003

Third Grade	1
Fourth Grade	1
Fifth Grade	1
Sixth Grade	2
Seventh Grade	2
Eighth Grade	5
Ninth Grade	5
Tenth Grade	9
Eleventh Grade	7
Twelfth Grade	67

Source: Author's e-government content analysis database
Note: The mean grade level is eleventh grade.

that agencies serving more of an educated clientele would gear their website to a higher level than those serving more poorly educated people. For example, research by Louis Tornatzky and K. Klein suggest a connection between mission and innovation.[20] As shown in table 3-7, however, agencies geared toward the less educated did not have lower grade-level readability levels. For example, corrections departments reported the highest percentage (83 percent) of websites written at the twelfth-grade level. This is problematic because its clientele is likely to fall well below the eighth-grade reading level of the typical American. Other agencies that had a high percentage of sites written at the twelfth-grade level were budget (81 percent), economic development (79 percent), elementary education (74 percent), housing (69 percent), health (69 percent), human services (67 percent), and taxation (46 percent).

Website readability levels vary significantly across individual states and the federal government. The state whose site was geared to the highest grade level in terms of mean readability was Utah with an average grade level of 11.7 across its websites. This was followed by Mississippi (11.5), Texas (11.5),

TABLE 3-7
Readability Level by Agency Type, 2003

	Education (%)	Human Services (%)	Health (%)	Housing (%)	Corrections (%)	Budget (%)	Taxation (%)	Economic Development (%)
Read at twelfth-grade level	74	67	69	69	83	81	46	79

Source: Author's e-government content analysis database

Virginia (11.5), Minnesota (11.4), Arkansas (11.4), and Idaho (11.4). The state whose websites were geared to the lowest grade level was Rhode Island, which had an average readability level of grade 10.1.

DISABILITY AND FOREIGN LANGUAGE ACCESS

A key measure of how e-government is progressing involves the manner in which technology serves populations with special needs. Part of the digital divide deals with whether all citizens share equally in the benefits of technology. According to Ben Shneiderman, "universal usability" is a key goal in technology assessment. From his standpoint, technology should enable "more than 90% of all households [to be] successful users of information and communications services at least once a week."[21] Anything that limits usability prevents citizens from taking full advantage of new technology.

Groups that stand out for particular concern in the public sector are those who are either visually or hearing-impaired, or those who do not speak English. One of the attendant virtues of electronic technology is the ability to tailor information to different kinds of people, depending on their specific circumstances. Rather than treating everyone identically, it is possible to develop sites that are accessible to the disabled and provide foreign language translation to non-English speakers.

For example, there is software available for the visually impaired that converts computer text to audio signals. This allows those who cannot see to hear the contents of a website read to them. Websites must be set up properly for this software to work, however. Images must have text labels for this reading software to be able to inform the visitor what the picture represents and tables must be set up in a clear and hierarchical manner. Further, for non-English speakers in the United States, websites can offer translation sites that provide Spanish and other languages to site visitors.

U.S. Commerce Department data indicate that "Internet access by people with disabilities is one-half that of people without disabilities." Only 22 percent of the disabled have access to the Internet, compared to 42 percent of Americans who do not have disabilities.[22] These kinds of disparities led Congress in 1998 to draft Section 508 of the U.S. Rehabilitation Act of 1973. This amendment requires federal agencies to comply with accessibility standards in all aspects of operations, including e-government. While not all parts of the site have to be accessible, major pages that are frequently visited are required to meet accessibility standards.

In 2003, we tested disability access by examining the actual accessibility of government websites. In the past, we had looked at whether sites displayed TTY (Text Telephone) or TDD (Telephonic Device for the Deaf) phone numbers, which allow hearing-impaired individuals to contact the agency by phone, and also whether sites provided text labels for graphics, or claimed generally that they were disability-accessible. This approach had the obvious disadvantage of not providing an actual test of accessibility, so in 2003 we used the online automated "Bobby" service at http://bobby.watchfire.com to test actual accessibility.

We used two different standards of website accessibility: compliance with the Priority Level One standards recommended by the World Wide Web Consortium (W3C) and compliance with Section 508 requirements. For each test, we entered the URL of the particular agency being evaluated and used this "Bobby" analysis to determine whether the website complied with either the W3C or the Section 508 guidelines. Sites were judged to be either in compliance or not in compliance based on the results of these two tests.

Thirty-three percent of state and federal sites satisfied the W3C standard of accessibility in 2003 and 24 percent met the guidelines for Section 508. Federal sites (47 percent) were more likely than state sites (33 percent) to meet the W3C standard of accessibility. There were few differences between states (24 percent) and federal sites (22 percent) when it came to meeting Section 508 accessibility standards, as measured by our Bobby analysis. These results are identical to those obtained by Jim Ellison, who found that 22 percent of federal sites passed the Bobby test.[23] Ironically, as websites have become more advanced and incorporated a variety of graphics and visual images, many of them have become less accessible to the visually impaired—unless they are accompanied by text labels that can be read by special software. The same is true for animation features that do not have information that is accessible through textual accompaniment.[24]

When looking at disability access for individual state sites, there is tremendous variation. The places doing the best job on disability access are North Dakota (84 percent of its sites are accessible using the W3C standard), Kansas (74 percent), New Hampshire (68 percent), and Texas (67 percent). The poorest states for W3C accessibility are New Jersey (none of its sites met the Bobby test), Mississippi (3 percent were accessible), and Iowa (10 percent compliance).

In 2002, we personally inspected disability access. The most common way government websites provided disability accessibility was through TTY/TDD phone lines, a feature that was available on 8 percent of sites. Eighteen percent offered text versions of their site or text labels for graphical images. Five percent claimed to be Bobby approved, and five percent were compliant with W3C or Section 508 regulations. These numbers were about the same as the previous year, suggesting that technological innovation in areas affecting the disabled has diffused fairly slowly.

Meanwhile, government sites are making slow but steady progress in providing foreign language accessibility. In our analysis, 13 percent of sites offer some sort of foreign language translation feature, up from 7 percent in 2002,

6 percent in 2001, and 4 percent in 2000. By "foreign language feature," we mean any accommodation to the non-English speaker, from a text translation into a different language to translating software available for free on the site that will translate pages into a language other than English.

According to federal statutes regarding election ballot access, if a community has a non-English-speaking population exceeding 5 percent, that local government is required to provide ballots in the native language of that minority population.[25] This requirement could be applied to government websites with respect to which foreign languages they place on their sites. If, based on Census data, the Spanish- or Chinese-speaking component of their state or city exceeds 5 percent, then that site should offer major documents and services in the relevant language for those groups. This would help to insure more equitable access to public sector websites for major groups within a community.

ADS, USER FEES, AND PREMIUM FEES

Budget resources are one of the major limiting factors in technological innovation. New technology costs money, and it takes jurisdictions with substantial revenues to develop electronic government. The dire budgetary straits of many governments creates incentives for the public sector to experiment with other revenue sources. Some of the options that have been proposed include having commercial ads on government websites, charging user fees (or convenience fees) to access specific services, or levying premium charges to enter particular website sections where business data are available.

We wanted to see to what extent these new revenue models have been put into use, and whether the patterns have changed over time. Despite the fiscal problems facing individual states and the national government, there have not been any significant shifts in the last few years (consistent with an incrementalism model). For example, there has been no increase in the use of ads to finance government websites in recent years. Whereas in 2002, 2 percent of sites had commercial advertisements on their sites—meaning nongovernmental corporate and group sponsorships—that figure was about 1 percent in 2003 (table 3-8). When defining an advertisement, we eliminated computer software available for free download (such as Adobe Acrobat Reader, Netscape Navigator, and Microsoft Internet Explorer) since they are necessary for viewing or accessing particular products or publications. Links to commercial products or services available for a fee were included as advertisements, as were banner, pop-up, and fly-by advertisements.

Examples of advertisements on the states' sites are the Arizona tourism site (Hilton and Four Seasons hotel packages), Colorado higher education ("Colorado Mentor" organization with the Xap Corporation and the "College

TABLE 3-8
Percentage of Sites with Ads, User Fees, and Premium Fees, 2001–2003

	2001 (%)	2002 (%)	2003 (%)
Ads	2	2	1
User Fees	2	2	3
Premium Fees	—	1	0.4

Source: Author's e-government content analysis database

Invest" organization), Colorado tourism ("Copper Mountain Resort," "Now This Is Colorado," "Buffalo Joe Whitewater Rafting," and "Grand-County .com"), New Mexico agriculture ("The Weather Channel"), and South Dakota economic development ("TravelSD.com").

Three percent of state and federal sites require user fees to access information and services, including archived databases of judicial opinions and up-to-the-minute legislative updates. This is about the same as in 2001 and 2002 (2 percent). Examples of states with user fees include Indiana's driver's license renewal and motor vehicle registration areas (a $3 charge in addition to the normal renew charges for processing the renewal through the online service, BMV express), Massachusetts wildlife license registration ($1 and $2 shipping and handling fees), Massachusetts conservation (an $8 reservation transaction charge for customers making reservations through "ReserveAmerica"), Arkansas portal ($6 for ordering death or birth certificates), Texas portal ($1 convenience fee for change of address and many other DMV services), and Wisconsin hunting (a $3 convenience fee for purchases that include any combination of licenses, permits, or applications).

Less than 1 percent of government websites require premium fees to access portions of the e-government site. By a premium fee, we mean financial charges that are required to access particular areas on the website, such as business services, access to databases, or viewing of up-to-the-minute legislation. This is not the same as a user fee for a single service. For example, we did not code as a fee the fact that some government services require payment to complete the transaction (a user fee). Rather, a charge was classified as a premium fee if a payment was required in order to enter a general area of the website or access a set of premium services. Subscription services were considered a premium fee if there was a cost associated with the subscription.

Examples of states with premium service areas include the Kansas secretary of state's website (Uniform Commercial Code filings may be done online but require a fee of $15 for the first ten pages of filing and $1 for each additional page; Financing Statement copies require an additional $1 per page); the Kansas legislature's website (you pay a $1 or $2 subscription fee for online bill-viewing and $50 per month for a "Lobbyist-in-a-Box" option to create pro-files of bills, monitor them, and alert the user to changes in the bill); Maine secretary of state (an annual fee of $75 for access to the Bureau of Motor Ve-hicles, special request services, Bureau of Corporations, Elections, and Com-missions, UCC searches, UCC filing, bulk databases, and Bureau of Identi-fication), and Arkansas Portal Information Network (an annual fee of $50 is required to access specific services such as database searching, workers com-pensation claims, and the Board of Nursing registry).

Restricted Areas

Some government websites have created restricted areas requiring a user-name and password to enter. These areas might afford access to government contract information or procurement bidding, or to a subscription- or business-services area that is password protected. We did not consider a section a re-stricted area if there was a registration requirement for a password just for information purposes, that is, sending free email notifications or free sub-scriptions to the visitor because these were not restrictions on a general area of the website. In addition, individual services that required a password for ex-ecution, such as income tax filing, were not considered to be a restricted area because the password involved that specific service, not a general area of the website. Sections providing access to state employee records that were pass-word protected were not coded as a restricted area because they contain in-formation that the general public does not have a right to see.

In 2003, 17 percent of sites had restricted areas. This was up from 6 per-cent the preceding year. Examples of states with restricted areas include part of South Dakota's portal page with access to Service Direct (an archive of all the state's forms that requires a login name and password), the Kansas secre-tary of state's website (which requires a connection to accessKansas Sub-scriber Services in order to conduct a Uniform Commercial Code search, file a UCC document, and view Kansas administrative regulations), the Texas at-torney general's website (requires a login and password to view subscriptions to weekly columns, senior alerts, consumer alerts, and law enforcement up-dates), the Utah Administrative Services "InnerWeb" (a center for state em-ployees that allows them to look at and change information on their W-4), and the Washington courts portal page (which requires attorneys to submit their bar number to have access to the dates of the court appearances in which they are an attorney of record and schedule appearances for the next week).

The danger of these restricted areas is that they create a "two class" society for e-government users. With access dependent on passwords and financial payment, such areas start to break down the free and open-access principles on which e-government previously has been based. In the long run, restricted access and premium-payment areas pose the danger that some people will gain greater access to government information and services than others. Placing either financial or structural limits on the ability of people to access government websites reduces Internet usage and discourages ordinary people from taking full advantage of the resources that are available online.

OVERALL E-GOVERNMENT RANKINGS

After looking at e-government performance in specific areas, we created a 0–100 point e-government index for each website and ranked the fifty states. On each state website, four points were awarded for the presence of the following twenty features: publications, databases, audio clips, video clips, foreign language access, not having ads, not having user fees, not having premium fees, not having restricted areas, W3C disability access, having privacy policies, security policies, allowing digital signatures on transactions, an option to pay via credit cards, email contact information, areas to post comments, option for email updates, allowing for personalization of the website, PDA or handheld-device accessibility, and readability levels below grade ten. These features provided a maximum of 80 points for particular websites.

Each site then qualified for up to 20 additional points based on the number of online services executable on that site (0 for no services, 1 point for one service, 2 points for two services, 3 points for three services, 4 points for four services, and so on up to a maximum of 20 points for twenty services or more). The e-government index therefore ran along a scale from 0 (having none of these features and no online services) to 100 (having all twenty features plus at least twenty online services). This total for each website was averaged across all of the state's websites to produce a 0–100 overall rating for that state. On average, in 2003, we assessed thirty-two government websites in each state across the executive, legislative, and judicial branches of government.

Table 3-9 shows that the top state in our ranking was Massachusetts. Looking across all of its websites on the dimensions we analyzed, it scored an average of 46.3. It was followed by Texas (43), Indiana (42.4), Tennessee (41.1), California (41.1), Michigan (40.7), Pennsylvania (40.5), New York (40.5), Florida (40.3), and Kentucky (40.0). The most poorly performing e-government states were Alaska (30.3), New Mexico (30.9), Nebraska (31.3), and Mississippi (31.5).

Federal sites were rated on the same dimensions as were the fifty states. An identical e-government index was devised that rated federal websites on contact information, publications, databases, portals, and number of online services. The unit of analysis was the individual federal agency. Overall, federal government websites did better than the states on our e-government index.

TABLE 3-9
Percentages for Overall State E-Government Performance, 2003

Massachusetts	46.3	Ohio	37.4
Texas	43.0	Minnesota	36.8
Indiana	42.4	Louisiana	36.6
Tennessee	41.1	North Dakota	36.4
California	41.1	Idaho	35.9
Michigan	40.7	Georgia	35.8
Pennsylvania	40.5	Nevada	35.7
New York	40.5	Rhode Island	35.3
Florida	40.3	Oregon	34.9
Kentucky	40.0	Iowa	34.6
Illinois	39.7	Wisconsin	34.2
Missouri	39.7	Arkansas	34.0
New Jersey	39.6	Oklahoma	33.2
South Dakota	39.5	Colorado	33.1
Arizona	39.1	Wyoming	33.0
Washington	38.6	West Virginia	32.7
Utah	38.1	South Carolina	32.7
Maryland	38.1	Montana	32.7
Virginia	38.1	Vermont	32.3
North Carolina	38.0	Hawaii	32.1
Kansas	38.0	Alabama	31.9
Connecticut	37.9	Mississippi	31.5
New Hampshire	37.6	Nebraska	31.3
Delaware	37.4	New Mexico	30.9
Maine	37.4	Alaska	30.3

Source: Author's e-government content analysis database

As shown in table 3-10, the sixty federal government sites clearly have made substantial digital progress. The best e-government performer was FirstGov, the United States national government portal, which scored an 84 out of 100. It was followed by the Federal Communications Commission (73), Social Security Administration (69), Internal Revenue Service (68), Library of Congress (68), Postal Service (68), Department of Treasury (64), Securities and Exchange Commission (64), Department of Housing and Urban Development (62), and the Consumer Product Safety Commission (57).

At the low end of the ratings were the various circuit courts of appeal and the U.S. Supreme Court. Eleven of the twelve lowest performers on our e-government index came in the federal judiciary. Their score ranged from a low of 24 (Eighth Circuit Court of Appeals) to 41 (Fifth Circuit). The new Homeland Security Department scored a 38, putting it in the lower third of federal agencies.

TABLE 3-10
Percentages for Overall Federal Agency E-Government Performance, 2003

FirstGov Portal	84	Federal Reserve	45
Federal Communications Commission	73	Congressional Budget Office	44
Social Security Admin.	69	NASA	44
Internal Revenue Service	68	Office of Management and Budget	44
Library of Congress	68	U.S. House of Representatives	42
U.S. Postal Service	68	5th Circuit Court of Appeals	41
Dept. of Treasury	64	Equal Employment Opportunity	41
Securities and Exchange Commission	64	Government Printing Office	41
Dept. of Housing and Urban Development	62	Dept. of Justice	40
Consumer Product Safety Division	57	Federal Deposit Insurance	40
Dept. of Agriculture	56	Nat'l Endowment for the Humanities	40
Dept. of Defense	56	Nat'l Transportation and Safety	40
General Services Admin.	56	Dept. of Homeland Security	38
Nat'l Science Foundation	56	Nat'l Labor Relations Board	38
Small Business Admin.	56	Dept. of the Interior	36
Dept. of State	54	U.S. Senate	36
Food and Drug Admin.	53	U.S. Supreme Court	36
White House	53	U.S. Trade Representative	36
Federal Trade Commission	52	11th Circuit Court of Appeals	34
Dept. of Health and Human Services	52	10th Circuit Court of Appeals	33
Dept. of Education	51	Federal Circuit Court of Appeals	33
Dept. of Transportation	51	3rd Circuit Court of Appeals	32
Dept. of Commerce	50	Nat'l Endowment for the Arts	32
Environmental Protection Agency	50	1st Circuit Court Appeals	29
Dept. of Energy	49	9th Circuit Court of Appeals	29
Dept. of Labor	49	7th Circuit Court of Appeals	28
General Accounting Office	47	2nd Circuit Court of Appeals	25
Dept. of Veterans Affairs	47	6th Circuit Court of Appeals	25
Federal Election Commission	46	4th Circuit Court of Appeals	24
Central Intelligence Agency	45	8th Circuit Court of Appeals	24

Source: Author's e-government content analysis database

We also looked at e-government performance in the seventy largest metropolitan areas in the United States. Paralleling our study of state and federal e-government, this analysis was designed to chart progress at the city level and to compare e-government development across levels of government. The goal was to see whether the diffusion of technological innovation depends at all on whether one is examining city, state, or national government. We used an identical scale to the other public sector levels. On average, we assessed 27.6 government websites in each city across executive, legislative, and judicial branches of government.

Table 3-11 shows that the top city in our ranking is Denver at 64.8 percent.

TABLE 3-11
Percentages for Overall City E-Government Performance, 2003

Denver	64.8	Los Angeles	33.4
Charlotte	57.4	El Paso	33.0
Boston	55.6	New York City	33.0
Louisville	53.5	San Antonio	32.5
Nashville	53.0	Columbus	32.1
Houston	49.3	Orlando	31.8
Salt Lake City	48.7	Las Vegas	31.4
Dallas	48.5	Birmingham	30.1
Oklahoma City	47.4	San Jose	30.0
Tucson	46.8	Chicago	29.9
Jacksonville	45.5	St. Louis	29.7
Kansas City	44.3	Fort Worth	29.6
Austin	44.1	Norfolk	29.5
Virginia Beach	43.0	Knoxville	29.4
Washington, D.C.	41.2	Providence	29.4
Phoenix	40.8	Sacramento	28.9
Memphis	40.0	Pittsburgh	28.6
San Diego	40.0	Long Beach	28.6
Milwaukee	39.8	Omaha	28.5
Richmond	38.8	Detroit	28.4
Tampa	38.4	Grand Rapids	28.3
New Orleans	38.2	Greensboro	28.0
San Francisco	38.0	Cleveland	27.8
Buffalo	37.4	Baltimore	27.6
Syracuse	36.5	Philadelphia	27.3
Fresno	36.2	West Palm Beach	27.0
Seattle	36.0	Albany	27.0
Albuquerque	35.7	Raleigh	26.7
Honolulu	35.6	Hartford	26.1
Cincinnati	35.5	Oakland	25.6
Minneapolis	35.2	Dayton	25.3
Rochester	34.8	Miami	25.1
Tulsa	34.6	Tacoma	23.9
Indianapolis	34.3	Atlanta	22.5
Portland	33.7	Greenville	22.2

Source: Author's e-government content analysis database

This means that the average website we analyzed for that city has nearly two-thirds of the features important for information availability, citizen access, and service delivery. Other cities that score well for e-government include Charlotte (57.4 percent), Boston (55.6 percent), Louisville (53.5 percent), Nashville (53 percent), Houston (49.3 percent), Salt Lake City (48.7 percent), Dallas (48.5 percent), Oklahoma City (47.4 percent), and Tucson (46.8 percent). The lowest ranked cities in our study included Greenville (22.2 percent), Atlanta (22.5 percent), Tacoma (23.9 percent), Miami (25.1 percent), and Dayton (25.3).

CITY-STATE-FEDERAL DIFFERENCES

Since we examined city, state, and federal government websites, we compared the three levels of government to see if there were differences in e-government progress by level. In general, we found there were significant differences. Federal sites are systematically ahead of the states and the largest cities, reflecting the greater financial resources at the national level and the federal government's ability to use economies of scale to spread technology costs over a larger geographic area.

On electronic services, for example, 68 percent of federal government sites offer some kind of services, compared to 44 percent of state sites and 48 percent of city sites (table 3-12). Ninety-five percent of federal government sites have databases, compared to only 79 percent of state sites and 41 percent of

TABLE 3-12
E-Government Performance by Level of Government, 2003

	Federal Sites (%)	State Sites (%)	City Sites (%)
Publications	100	98	79
Database	95	79	41
Foreign Language	40	12	16
Ads	2	1	1
Premium Fees	0	0	0
Restricted Areas	30	17	4
User Fees	0	3	7
Privacy Policy	75	53	41
Security Policy	62	36	27
W3C Disability Access	47	33	20
Services	68	44	48
Credit Cards	32	19	29
Email Addresses	93	90	71
Comment	52	23	35
Email Updates	32	11	8
Website Personalization	5	2	4
PDA Accessibility	0	1	0
Reads at Twelfth-Grade Level	63	68	70

Source: Author's e-government content analysis database

city sites. The federal government also has made greater progress in the area of disability access (47 percent of sites offer some form of disability access compared to 33 percent of state sites and 20 percent of city sites). Seventy-five percent of federal sites offer a privacy policy, compared to 53 percent of state government websites and 41 percent of city websites. Sixty-two percent of federal sites have a visible, online security policy, compared to 36 percent of the state sites and 27 percent of the city sites.[26]

These differences show the importance of the organizational, fiscal, and political forces noted in the previous chapter. The federal government, which has the greatest financial resources and broadest eonomy of scale, has made the greatest e-government progress. It has done the best at building online services, serving special populations such as the disabled, and reassuring the public about privacy and security. This is consistent with the logic of our conceptual model.

DIFFERENCES BY BRANCH OF GOVERNMENT

We also examined differences in e-government performance by branch of government. Given the differing missions of executive agencies, the legislative branch, and the courts, one would expect some variation. In fact, there are differences across the branches (table 3-13). Executive sites are the ones most likely to have privacy policies, security policies, and online services. This is in line with their public outreach and service role. Legislative sites, in contrast, have the greatest percentage of databases, audio clips, and video clips. In general, however, legislative and judicial pages lag behind executive pages in providing online services and have made considerably less progress on many specific indicators.

DIFFERENCES BY AGENCY TYPE

A number of academics have suggested that e-government performance varies by agency type. For example, Daniel Bugler, Stuart Bretschneider, and Charles Hinnant have argued that those geared toward outside constituencies will be the ones most likely to incorporate new technology.[27] Because they are externally driven, these organizations will have the strongest incentives to innovate.

As shown in table 3-14, this is exactly what we find in terms of the match between agency mission and specific e-government performance. For example, health departments were the most likely to have databases, while budget departments, which are internally driven, were the least likely. Economic development sites (which are typically geared toward business interests) were

TABLE 3-13
E-Government Performance by Branch of Government, 2003

	Executive (%)	Legislative (%)	Judicial (%)
Publications	98	98	96
Database	78	87	82
Audio Clip	6	36	6
Video Clip	8	28	12
Foreign Language	14	4	10
Ads	2	0	0
Premium Fees	0	2	0
Restricted Areas	17	9	16
User Fees	3	0	0
Privacy Policy	56	29	38
Security Policy	39	22	18
WC3 Disability Access	33	31	37
Section 508 Access	23	22	34
Services	47	12	25
Digital Signature	0	0	0
Credit Cards	19	4	9
Email Addresses	91	90	82
Comment	24	17	21
Email Updates	13	7	4
Website Personalization	1	2	0
PDA Accessibility	0	3	0

Source: Author's e-government content analysis database

the most likely to offer online services, while budget departments were the least likely. Health, housing, human services, and education agencies were the most likely to offer foreign language translation, which is consistent with the external nature of their mission to serve the poor, who, as a population, are more likely to contain non-English speakers. Economic development sites were the least likely to be accessible to the disabled.

CONCLUSION

To summarize, we find that some progress has been made from 2000 to 2003 in e-government information and services. Many city, state, and federal agencies have moved from the billboard stage to at least the partial service-delivery stage. Some even have developed fully integrated portals that offer a range of online services. In general, however, while there have been improvements in access to publications, databases, and the creation of portals, many govern-

TABLE 3-14
E-Government Performance by Agency Type, 2003

	Education (%)	Human Services (%)	Health (%)	Housing (%)	Corrections (%)	Budget (%)	Taxation (%)	Economic Development (%)
Publication	98	100	100	100	100	100	100	100
Database	92	86	98	78	88	73	88	88
Audio Clip	6	7	6	0	0	3	0	2
Video Clip	14	2	2	6	2	3	0	12
Foreign Language	16	19	31	31	2	3	4	9
Ads	2	0	0	0	0	0	2	2
Premium Fees	0	0	0	0	0	0	2	0
Restricted Areas	32	21	18	22	17	19	25	12
User Fees	2	2	4	6	4	0	6	2
Privacy Policy	54	51	67	58	52	51	73	58
Security Policy	38	30	45	32	35	35	52	32
WC3 Disability Access	30	26	33	28	25	46	33	19
Section 508 Access	18	19	28	22	21	38	27	9
Services	42	37	47	39	29	24	39	46
Digital Signature	0	0	2	0	0	3	4	0
Credit Cards	4	12	28	11	10	5	62	5
Email Addresses	96	84	94	97	90	86	92	93
Comment	24	26	24	28	17	16	29	28
Email Updates	14	0	10	11	6	0	21	28
Website Personalization	4	0	0	0	0	0	0	5
PDA Accessability	0	0	0	0	0	0	0	0

Source: Author's e-government content analysis database

ment websites are not offering much in the way of online services. A combination of budgetary limitations plus organizational and political factors has impeded the expansion of e-government. Virtually none of the sites examined meet the fourth e-government stage of interactive democracy, one of the indicators of transformational change. As shown later in this study, most sites do not offer much in the way of democratic outreach or provide accountability-enhancing devices that improve the functioning of the political system.

Part of the problem is that government officials are oriented toward service delivery rather than system transformation. They do not see technology as a tool to revolutionize the system and enhance political participation, but as a way to deliver services to the middle class. It is a worldview that limits the ability of the public sector to take full advantage of the technology that already is available. This inability to grasp the broader significance of technology limits the opportunity of the public sector to innovate and create new models of governance.

As it stands right now, only a moderate number of sites offer access to the disabled or non-English speakers. Many do not have visible security or privacy policies, or do not devote adequate attention to safeguarding the rights of visitors, although these efforts are up over those of preceding years. These problems limit the breadth of the change that is possible through the Internet and the ability of government officials to make meaningful improvements in organizational performance.

Progress has been made in state and federal governments creating websites that have more uniform, integrated, and standardized navigational features. This is crucial because Internet information and service delivery often has had weak consistency across websites. Government agencies guard their autonomy very carefully, and it has taken a while to get agencies to work together to make citizen tasks easier to undertake. Common navigational systems help the average citizen make use of the wealth of material that is online.

In looking toward the future, states need fully functioning portals that serve as gateways to their various sites. A number of states have adopted "one-stop" portals that provide a common location for citizen and business services. This is a tremendous help to citizens interested in making use of online resources, and offers the potential for improved citizen access to government websites.

Having created governmental portals, city, state, and federal governments need to publicize the existence of these service portals to the average citizen. According to a 2000 national survey conducted by Hart and Teeter for the Council for Excellence in Government, only 54 percent of Americans have logged onto a federal government website. While some of this access problem reflects a lack of computer availability, some citizens clearly need to be educated as to the existence of online government services. Marketing tools such as placing the portal address on state documents, putting the address on vehicle license plates (as done by the state of Pennsylvania, for example), and using televised public service announcements would help the average citizen learn where to go to make use of existing information and services.

Judging from the content of thousands of American city, state, and national government websites examined in 2000, 2001, 2002, and 2003, there is little empirical evidence to support a model of transformational change defined as a "complete change in character, condition" of service delivery in the United States.[28] While there have been rapid advances in putting publications and contact information online (features of the billboard stage), there has been little progress with respect to service delivery or security and privacy, and little change in regard to patterns of commercial advertising, credit card transactions, and disability accessibility. While electronic technology has the potential to alter future government performance, it has not produced dramatic changes in the short run.

CHAPTER 4

Explaining E-Government Performance

IT IS NOT ENOUGH to describe the scope of e-government change. Rather, observers must explain why the speed and breadth of technological innovation has shifted in the manner it has. In the opening chapters, I discussed a conceptual model based on organizational, fiscal, and political dimensions of technological change, noting that the pace of policy innovation varies with organizationals attributes, the level of fiscal resources, and political determinants, such as partisanship and group demands.

In this chapter, I apply this model to state e-government performance. Using several measures of e-government across the fifty American states, I examine the factors that help us understand aggregate, state-level performance. Who do some states perform better with respect to e-government than do others? Basically, I argue that legislative professionalism and fiscal capacity are important predictors of electronic performance. There is, however, considerable variation in explanatory factors depending on the e-government aspect that is being explored. Professionalism predicts the number of online services, but fiscal resources are more important for overall e-government ranking and the quality of privacy policies. There is no link between e-government performance and either privatization or the size of state budget deficits.

AGGREGATE ANALYSIS

Befitting the data analysis reported in the last chapter, I focus on online content for my analysis of e-government. Because there is no agreed-upon standard for the assessment of e-government performance, I study a number of different indicators from online services and readability to interactivity and privacy. I use these features because they represent important aspects of stage three and four e-government. Rather than picking one standard (such as online services) and making it the center of the evaluation, I employ a multidimensional perspective that relies on different indicators. This allows me to look at a variety of features and see what factors explain each of them.

In addition, I model overall e-government through an index that encapsulates services, interactivity, accessibility, and democratic capacity. This offers the advantage of presenting an overarching perspective on the topic, as opposed to one based on specific aspects of electronic government. It relies on the logic of stage four e-government and incorporates a number of different features that are important for information availability, online services, citizen accessibility, and democratic capacity.

By studying various features of online government, one can see if factors such as organizational features, fiscal resources, and political determinants have the same impact on every aspect of e-government or whether some of these factors are more important in some areas than others. Later in this chapter, I demonstrate that organizational, economic, and political factors vary considerably in their explanatory power, depending on what aspect of e-government is being explored; there is no single factor that explains every aspect of digital government or digital democracy.

In this analysis, I employ an aggregate approach to explaining e-government. My model focuses on states as the unit of analysis and utilizes the state-level performance indicators presented in the last chapter. Given the variation that exists across the fifty states, this level of government make an ideal laboratory in which to test different explanations. I focus on the aggregate state level because that is where much of e-government is being implemented.[1] States are spending millions of dollars on digital government and there are specific state-level indicators of organizational, fiscal, and political characteristics.

For the dependent variable, I look at the number and breadth of online services, website readability, quality of privacy policies, and overall e-government performance. These variables tap different dimensions of performance, from service and accessibility to interactivity and privacy protections. The service dimension is measured in two different ways. First, using the 2003 state e-government data set, we calculated the number of online services accessible through each website and averaged that across all sites within that state. This gives a measure of the depth of electronic service delivery. Second, we measured the breadth of service availability through the percentage of websites within each state that offer any services. Both depth and breadth of online services are important indicators of e-government progress.

Readability is measured through the average literacy level of websites within each state. Researchers evaluated readability by employing the Flesch-Kincaid test, a standard reading-tool evaluator used by the U.S. Department of Defense. As previously mentioned, the test computes reading level by dividing the average sentence length by the average number of syllables per word. This measure varies from first-grade to college-level literacy, based on the average of all state websites included in our data.

The quality of privacy policy is investigated through a 0 to 4 point scale for the presence or absence of the following four dimensions: whether the policy prohibits commercial marketing of visitor information (meaning it does not give, sell, or rent visitor information to third parties), whether the site prohibits creation of cookies or individual profiles of visitors, whether the site prohibits sharing personal information without prior user consent, and whether the site says it can share personal information with legal authorities or law enforcement. Each of these items was coded a 0 for no and a 1 for yes. The quality index then was an additive scale measuring the presence of zero to four privacy protections.

Overall e-government performance is a 0 to 100 point index that includes many different aspects of electronic activity. As discussed in the last chapter, 4 points are awarded each website for the following twenty features: publications, databases, audio clips, video clips, foreign language access, not having ads, not having user fees, not having premium fees, not having restricted areas, W3C disability access, having privacy policies, security policies, allowing digital signatures on transactions, an option to pay via credit card, email contact information, areas to post comments, option for email updates, allowing for personalization of the website, PDA or handheld device accessibility, and readability levels below grade ten. These features provide a maximum of 80 points for particular websites.

Each site then qualifies for up to 20 additional points based on the number of online services executable on that site (0 for no services, 1 point for one service, 2 points for two services, 3 points for three services, 4 points for four services, and so on up to a maximum of 20 points for twenty services or more). The e-government index therefore runs along a scale from 0 (having none of these features and no online services) to 100 (having all twenty features plus at least twenty online services). This total for each website is averaged across all of the state's websites to produce a 0 to 100 overall rating for that state. On average, in 2003 we assessed approximately thirty-two websites in each state across the executive, legislative, and judicial branches of government.

For the independent variables in this study, we looked at several explanations derived from the conceptual model presented earlier: organizational, fiscal, and political determinants of state performance. The organizational factors included a measure of interest group lobbying. Utilizing a survey undertaken by the Council of State Government in January 2002, we compiled the number of registered lobbyists in each state.[2] This measure is designed to tap the extent of group conflict within each state. Group conflict is a factor that can speed or slow the bureaucracy's capacity for technological innovation. In addition, we used education level within each state, measured by the percentage of the population that has graduated with a college degree. This is based on the 2000 U.S. Census and taps the degree of training within each state's labor force. Finally, we relied upon a standard measure of legislative professionalism developed by Peverill Squire.[3] Professionalism is a measure of organizational capacity that has been shown to be related to a number of dimensions of state policy innovation.[4] Squire's measure is based on the staffing and salary of members of state legislatures.

The fiscal measure was per capita personal income in 2000. Drawing data from the U.S. Census, we use the average number for each state.[5] This figure ranges from a low of $21,017 in Mississippi to a high of $41,392 in Connecticut, and it measures the state's financial capacity for undertaking policy innovations. This indicator is designed to tap the relative impact of wealth and poverty on a state's ability to implement new technology in the public sector.

The political determinants included three different items measuring party competition and citizen demand. For party competition, we rely on the percentage of Democrats in the state's House of Representatives.[6] Partisan competition has been linked to a large number of policy areas.[7] Depending on the issue, having more Democrats or more Republicans is important for funding specific initiatives. In terms of citizen demand, we looked at the percentage of each state's citizens who say they use the Internet and the percentage of a state's government websites where citizens can communicate and transact online business with public sector agencies.[8] These indicators are designed to tap the extent to which citizens have the technological means to demand e-government performance in each state.

Number and Breadth of Online Services

One of the most important aspects of e-government is the number and breadth of online services. States vary widely along both of these dimensions. In 2003, for example, Massachusetts had the largest average number of services online (25.4) across its sites, while Alaska had the smallest average (0.2). In terms of the percentage of state agencies offering online services, Massachusetts had the highest percentage (73 percent), compared to Wyoming and Alaska, which had the lowest (18 percent). The correlation between depth and breadth of online services in 2003 was 0.47, indicating an association but incomplete overlap between the two measures.

To examine the factors that explain state rank, we undertook an ordinary least squares regression of the variables noted in the last section on the number of online services and the percentage of state agencies providing online services (table 4-1). The results demonstrate that different factors explain depth and breadth of service offerings. For number of services, the most important determinant was legislative professionalism. The more professional the legislature, the more online services were offered by those states. There were no significant relationships between fiscal capacity, party competition, or political demand for e-government. This suggests that states have the ability to innovate on technology beyond their wealth or party composition.

In contrast, the best predictor of the *percentage* of state agencies offering online services was the state's personal income per capita. States with higher personal incomes were more likely to have a larger number of agencies providing online services. Those without a strong fiscal base were not in a strong position to have widespread online service delivery across a large number of agencies.

TABLE 4-1

The Impact of Organizational, Fiscal, and Political Factors on the Number and Percentage of Online Services in the Fifty States, 2003

	Number of Online Services	Percentage of State Agencies Offering Online Services
Number of Group Lobbyists	−.00 (.00)	.00 (.00)
Percent College Graduates	.03 (.17)	−.60 (.70)
Legislative Professionalism	8.53 (3.90)*	3.51 (15.96)
State Personal Income	.00 (.00)	.00 (.00)*
Percent Democrats in Legislature	.06 (.04)	−.01 (.15)
Citizen Internet Usage	.05 (.12)	−.46 (.50)
Percent of Websites Where Citizens Can Communicate with the Agency	−.03 (.04)	.11 (.18)
Constant	−7.06 (6.11)	37.29 (24.98)
Adjusted R Square	.19	.02
F	2.64*	1.14
N	50	50

$*p < .05; **p < .01; ***p < .001$
Source: Author's analysis
Note: The numbers are the unstandaradized least squares regression coefficients, with the standard error in parentheses. The number of asterisks indicates the level of statistical significance. Tolerance statistics show no multicollinearity problem in the model.

These results demonstrate that on the breadth of agency service offerings, fiscal capacity is a determining force, not legislative professionalism, party competition, or citizen demands for e-government services. It matters less how professional the legislature is, what the party composition is, or whether citizens are in a position to request or need more services. None of these factors were linked to the percentage of agencies offering services.

WEBSITE READABILITY

Website readability is a measure of the accessibility of government sites. Citizens need websites that are legible at a level they can comprehend. According to national statistics, about half of the American population reads at the eighth-grade level or lower. If government websites are written at too high of a level, then it is difficult for a wide range of citizens to comprehend online material.

To see what factors explain government website readability, we regressed organizational, fiscal, and political characteristics on the average grade level of a state's websites. As shown in table 4-2, there were three factors that were

TABLE 4-2
The Impact of Organizational, Fiscal, and Political Factors on Website Readability in the Fifty States, 2003

	Website Readability
Number of Group Lobbyists	.00 (.00)
Percent College Graduates	.04 (.02)*
Legislative Professionalism	.10 (.39)
State Personal Income	−.00 (.00)
Percent Democrats in Legislature	−.01 (.00)*
Citizen Internet Usage	−.03 (.01)**
Percent of Websites Where Citizens Can Communicate with the Agency	.00 (.00)
Constant	12.86 (.61)***
Adjusted R Square	.18
F	2.53*
N	50

*$p < .05$; **$p < .01$; ***$p < .001$

Source: Author's analysis

Note: The numbers are the unstandaradized least squares regression coefficients, with the standard error in parentheses. The number of asterisks indicates the level of statistical significance. Tolerance statistics show no multicollinearity problem in the model.

significantly related to readability level: the percentage of college graduates within each state, the percentage of Democrats in the legislature, and the percentage of citizens who use the Internet in each state. The more college graduates there were, the fewer Democrats in the legislature, and the fewer Internet users there were in the state, the higher the grade level of readability of public sector websites.

These results are interesting because they demonstrate that organizational and political demands condition the way in which websites are designed. Even beyond fiscal capacity, states where there are social and public pressures to keep readability to a low grade level actually deliver lower grade level readability than areas where the pressures move in the other direction. This suggests that website designers are sensitive to the larger environment in which they work, regardless of the state's wealth or level of professionalism.

QUALITY OF PRIVACY POLICY

This analysis also examined the quality of privacy policies within each state. We used a 0 to 4 point privacy index based on four dimensions (prohibition of commercial marketing of visitor information, prohibition of cookies or individual profiles of visitors, prohibition of sharing personal information without prior user consent, and whether the site says it can share personal information with legal authorities or law enforcement). States scoring a 4 would have the highest quality privacy policies since they meet all four of the conditions, while those with a 0 have the lowest quality because they fail to meet any of the four privacy requirements.

Table 4-3 presents the results of the regression model. The two factors that best explain the quality of privacy policy within each state are personal income and citizen ability to communicate with government websites. The higher incomes there are in a state and the more ability citizens have to communicate with public sector sites, the higher the quality of the state's privacy policy. This demonstrates that fiscal capacity and political demands affect privacy policy, even more so than organizational features or party competition (neither of which were statistically significant).

The communications capability result is noteworthy because it shows that capacity is linked to outcomes. Privacy is an issue about which citizens repeatedly have expressed concern in public opinion surveys. The average person worries whether information provided online or websites accessed will

TABLE 4-3

The Impact of Organizational, Fiscal, and Political Factors on the Quality of Privacy Policy in the Fifty States, 2003

	Quality of Privacy Policy
Number of Group Lobbyists	.00 (.00)
Percent College Graduates	−.07 (.04)
Legislative Professionalism	.90 (.89)
State Personal Income	.00 (.00)*
Percent Democrats in Legislature	.01 (.01)
Citizen Internet Usage	.04 (.03)
Percent of Websites Where Citizens Can Communicate with the Agency	.02 (.01)*
Constant	−3.28 (1.39)*
Adjusted R Square	.26
F	3.44**
N	50

$*p < .05; **p < .01; ***p < .001$
Source: Author's analysis
Note: The numbers are the unstandaradized least squares regression coefficients, with the standard error in parentheses. The number of asterisks indicates the level of statistical significance. Tolerance statistics show no multicollinearity problem in the model.

lead to an invasion of personal privacy. When websites are set up in a manner that facilitates citizen communication, those states respond with strong privacy protections.

OVERALL PERFORMANCE

The results presented so far demonstrate that organizational, fiscal, and political factors vary considerably in their ability to explain different aspects of e-government performance. Legislative professionalism mattered the most for a number of services, while state fiscal capacity was more important for the percentage of a state's sites offering services, and fiscal capacity and citizen demands affect the quality of the state's privacy policies.

It remains to be seen, however, what is most important in assessing overall e-government performance. The previous analysis looks at e-government along several specific dimensions. This is important to see what factors are linked to particular technological improvements. But what happens when one incorporates a variety of different features into an overall scale? We use state rank and state numeric scores, respectively, to assess overall e-government performance. State rank runs from 1 (the highest ranked state) to 50 (the lowest ranked state). State numeric performance runs from a low of 0 to a high of 100 points. Features assessed include information availability, online services, accessibility, interactivity, and democratic capacity.

In order to examine what factors explained state rank, we undertook an ordinary least squares regression model that looked at the impact of various state factors. As shown in table 4-4, the only factor associated with state rank or numeric performance was state fiscal capacity. The wealthier the state, the higher that state tended to rank and score on overall e-government. There was no association on either measure with organizational factors, political demands, or party competition. This suggests that, in the end, when a variety of indicators are incorporated into the analysis, money matters most. Without sufficient financial resources, it is difficult to innovate and incorporate technology into the public sector.

THE IMPACT OF PRIVATIZATION AND BUDGET DEFICITS

Beyond the factors cited above, there are two specific characteristics that have attracted attention for their possible link to e-government performance: privatization and the presence of budget deficits. Some states have started to privatize public sector functions. For example, a 2002 Council of State Government survey of state budget directors found that of the thirty-nine states that responded to a question about privatization, 31 percent said they had in-

TABLE 4-4

The Impact of Organizational, Fiscal, and Political Factors on Overall
E-Government Performance in the Fifty States, 2003

	State Rank		State Numerical Performance	
Number of Group Lobbyists	−.00	(.00)	.00	(.00)
Percent College Graduates	1.04	(.71)	−.25	(.18)
Legislative Professionalism	−25.52	(16.26)	5.68	(3.99)
State Personal Income	−.002	(.001)*	.00	(.00)*
Percent Democrats in Legislature	.11	(.16)	−.02	(.04)
Citizen Internet Usage	.10	(.51)	−.03	(.12)
Percent of Websites Where Citizens Can Communicate with the Agency	−.20	(.18)	.04	(.04)
Constant	51.14	(25.45)*	29.47	(6.25)***
Adjusted R Square	.18		.19	
F	2.52*		2.65*	
N	50		50	

*$p < .05$; **$p < .01$; ***$p < .001$

Source: Author's analysis

Note: The numbers are the unstandaradized least squares regression coefficients, with the standard error in parentheses. The number of asterisks indicates the level of statistical significance. Tolerance statistics show no multicollinearity problem in the model.

creased the amount of privatization in the previous five years, 13 percent had decreased it, and 56 percent said it had stayed about the same.[9] Around this time, a 2002 survey by the National Conference of State Legislatures found 34 percent of the fifty states had no budget deficits, 32 percent had deficits under 5 percent of the general fund, and 34 percent showed deficits of five percent or greater.[10]

There was not, however, a significant relationship between state website rank and either privatization or budget deficits. States that said they were increasing the amount of privatization had an average rank of 25.2, compared to 24.9 for states that were keeping privatization levels the same and 29.4 that said they were decreasing privatization. This indicates that those increasing privatization had a better rank than those decreasing it, but the difference was not statistically significant. Similarly, states with no deficit averaged 24.9 on their e-government ranking, while those with a deficit under 5 percent had an average rank of 25.6 and those with a deficit 5 percent or greater of their general fund had an average e-government rank of 26.0. These differences in means were not statistically significant.

To see how these qualities related to various e-government indicators, we converted privatization into a 3-point scale where 1 represents decreased use,

TABLE 4-5

Correlations between Privatization, Budget Deficits, and Various State
E-Government Indicators, 2003

	Privatization	Budget Deficits
State Rank	−.07	.17
State Numeric Performance	.06	−.20
Number of Online Services	−.06	−.04
Percent of Agencies Providing Services	.00	−.15
Website Readability	−.19	.03
Quality of Privacy Policy	−.11	−.04
N	39	50

Source: Author's analysis
Note: None of these bivariate correlations are statistically significant at the .05 level or below.

2 refers to keeping privatization levels the same, and three indicates increased privatization. For budget deficits, I employed a percentage scale reflecting the budget's percentage of the state's general fund. This ran from a low of 0 percent deficit to a high of 20.6 percent in Alaska.

Table 4-5 reports bivariate correlations between privatization, budget deficits, and various indicators of e-government performance. There was no association between privatization and e-government performance. The feature showing the strongest correlation was website readability ($r = -.19$), indicating that states making increased use of privatization tended to have lower grade-level readability. But neither this association nor any of the others were statistically significant.

The same was true for budget deficits. There was a positive association between deficits and state rank, which indicates that states with larger percentage deficits had lower ranked websites. But the correlation was not large enough to be considered statistically significant.

CONCLUSION

To summarize, the results reported in this chapter demonstrate that different areas of e-government are explained by organizational, fiscal, and political qualities. Legislative professionalism predicts the number of online services, but fiscal resources are more important for overall e-government ranking and the quality of privacy policies. Citizen demands are most central with regard to website readability. There is no association between e-government performance and either privatization or state budget deficits.

Speaking more generally, this chapter demonstrates that organizational, fiscal, and political factors are important for e-government, but money is most crucial in terms of overall performance. States with the financial means to fund digital government are the ones that have earned the highest scores and received the highest ranks. Party competition does not matter as much; neither do organizational qualities affect overall performance.

In their study of technological change, the Digital State Survey emphasized political leadership as a crucial factor in policy innovation. What is most telling with regard to leadership is the ability to draw financial resources together and overcome bureaucratic intransigence, which weakens the ability of the public sector to incorporate technology into its mission. These skills are intangible but critical to the successful adoption of technology.

In comparing IT to other topics, it is important to recognize its differences. Unlike policy areas such as education and health care that are high-profile and redistributive in nature, electronic governance represents an example of what Theodore Lowi described as a "constituent policy" domain.[11] It is a low visibility area in which dynamics are driven by internal rather than external considerations.[12] The subject generally attracts little public or media attention. Unless there is a contracting scandal, e-government is not a topic that attracts much press coverage.

This lack of general visibility frees bureaucrats and administrators to innovate as they see fit, away from the glare of the public spotlight. School superintendents who want to revamp schools or alter the curriculum recognize there will be considerable public scrutiny and media coverage of their activities. Such interest is not likely to occur with respect to digital government, which gives administrators more autonomy than they would have in dealing with other types of policy issues.

Unfortunately, the downside of this lack of attention is difficulty in marshalling financial resources to pay for technology innovation. From this study, it is clear that money is crucial to making progress in the area of e-government. New technology requires money, as does meeting the needs of special populations or incorporating interactive features into government websites. Without public interest or media coverage, it is difficult to assemble the resources required for technological change. Administrators are left to their own devices, unless they happen to reside in areas where politicians have a personal interest in technology. Short of that advantage, many states suffer because of their inability to marshal the resources necessary to introduce technology into the public sector.

CHAPTER 5

The Case of Online Tax Filing

NOTHING ILLUSTRATES the opportunities and challenges of technological change more than the subject of electronic services. As pointed out in previous chapters, public officials have placed a high priority on getting services online. Consistent with a technocratic vision aimed at the business community and middle class, the goal of many city, state, and federal planners is to allow Internet users to register cars, file for business permits, and pay taxes online, among other features. Professional associations of IT experts are pushing for these options, and elected officials have embraced Internet service delivery as a long-term objective.

There are several reasons for this focus on service delivery. The first is budgetary in nature. Services are an expensive part of government agencies. This is the feature of the public sector mission that requires office space, financial resources, and staff. Since services are expensive to provide in person and over the telephone, government officials hope that electronic delivery will lower the costs of service delivery, thereby saving the government money. If usage levels rise and people make more extensive use of online services, government agencies can redirect staff, office space, and financial resources from in-person service delivery to electronic services. This solves the two systems problem and frees money from the bricks-and-mortar government for digital technology.

Second, there is a political imperative for officials to use a service orientation so as to build a constituency for change. Most government innovations—technological or otherwise—require an external constituency. Someone from outside the public sector must be interested in and supportive for there to be a significant expenditure of government resources. This support could come from private groups, business associations, or organized citizenry. In the case of e-government, public officials have sought to build support in two ways. They have appealed to the business community by making available contracts for the development and operation of service applications. In addition, they have emphasized service delivery to the middle class, educated, and affluent members of society because these are the individuals most likely to use the Internet and thereby benefit from online service delivery. Few of the alternative e-government visions that are available, such as interactive democracy or participatory democracy, offer the external support that is present in the area of service delivery. With the exception of some advocacy organizations, there is not a well-organized body of citizens pushing the public sector to embrace the Internet as a tool for direct democracy.

Finally, the emphasis on service delivery relates to broader visions about social change. Of all of the future e-government scenarios, a service-delivery focus is the one that is least threatening to the status quo. Options such as interactive democracy, participatory democracy, or direct democracy are possible ways to transform the political system and society as a whole. Grassroots organizations would be empowered by these models of democracy. Those who lack political power or financial resources could use technology to contest government policy.

Those visions of e-government clearly challenge major societal interests, however, so they have not been emphasized nearly as much by public officials as options based on electronic service delivery. Rather than pursue a vision that is threatening to the social and political status quo, e-government officials have preferred a technocratic vision emphasizing service to the business community and middle-class Internet users. These are the people who vote and contribute campaign monies so it is a way to build longer-term political support for technology initiatives.

In this chapter, I present a case study of online tax filing in order to illustrate broader dynamics about technological change. I discuss online tax filing at both the federal and state levels to show how it emerged, what its costs are, and how successful it has been in terms of providing electronic services. Using data collected from the federal government and a number of states about their experiences with online tax filing, I show that less than half of Americans are using online tax filing. Although citizen satisfaction with this electronic service is quite high, usage levels are growing below government projections and appear to be more consistent with models of incremental than transformational change. I close the chapter by describing the factors that are constraining the growth of online tax filing.

ONLINE TAX FILING

One of the most popular e-government services is tax filing. Citizens who are comfortable with the Internet like online tax service because it is convenient, speedy, and reliable. Unlike the postal system, which occasionally loses parcels or has lengthy delivery delays, electronic transmission of tax materials is efficient and effective. In addition, for those awaiting tax refunds due to overpayments, both state and national governments advertise that people who file electronically will receive refunds in less than two weeks, which is twice as quick as those who send in paper copies. Electronic filing furthermore provides new tools for those who have a physical disability. On the IRS site, for example, the disabled can access "talking tax forms" developed by the agency that work with speech-generation software to file taxes online.[1]

In general, the public is very enthusiastic about online tax filing. Federal electronic taxpayers receive an IRS acknowledgment that their forms have been tabulated and that their math adds up correctly. This increases confidence on the part of taxpayers that financial information has been filed correctly. If a proposal from President George W. Bush is adopted, the federal government would give electronic filers an extra ten days beyond the current deadline to submit national returns.[2] Since the federal government has set a goal of having 80 percent of taxpayers filing tax returns online by 2007, this proposal for extra time is designed to provide an additional incentive for electronic filing.

Government officials like online filing because it reduces errors when public sector employees enter information from paper returns. It also cuts the flow of paper into filing centers. With millions of individuals and companies currently filing tax returns, the cost of storing and processing paper forms is very high. The government must maintain a very expensive staff and infrastructure to collect taxes in the United States. Since public officials would like to reduce the costs associated with tax collection, they have turned to online filing as a way to achieve this goal.

Furthermore, public officials like online filing because it generates revenue for other activities. Unlike alternative services that offer vital functions but do not generate much money, tax filing is the primary mechanism by which the government collects revenues for a wide range of services. In this situation, it is not surprising that public officials have made online tax filing one of their highest e-government priorities.

Varieties of Online Filing

Non-paper tax filing emerged over the last decade as technology was developed that made it possible for taxpayers to file returns through phones and then electronically. In 1993, the federal government started a service called Federal State Electronic Filing (FSEF). This program allowed citizens in states that authorized this service to file taxes with the federal government that included both federal and state income tax payments. Following its receipt of citizen tax returns and payments, the federal government forwards the relevant material to the respective states.

In Colorado, for example, citizen usage of this program increased substantially from 2,316 in 1993 (the first year following authorization) to 448,517 in 2000. Missouri saw a similar explosion in its FSEF program. Whereas 47,000 utilized it in 1995, the first year that the option was offered, the number had grown to 705,000 by 2001.[3] These are just a few of the states that have seen tremendous increases in the number of citizens filing their tax returns electronically.

With the clear popularity of FSEF and its ease of use, many states saw the convenience of online filing and developed their own options for electronic tax service delivery. TeleFile was one possibility, whereby citizens could use the telephone to transmit information. Under this program, taxpayers employed their telephone keypad to send income and tax information to the government, and thereby satisfy the requirement of filing tax returns.

In general, however, this option proved not to be very popular. In Colorado, for example, 42,519 people accessed TeleFile in 1998 (the first year it was authorized in that state), but that number dropped to 36,104 by 2000. TeleFile was more popular in some states, such as Illinois, as its users went from 14,008 in 1994 to 147,014 in 2001. But even there, TeleFile was less prevalent than other electronic mechanisms.

The most popular state-level tax option has been E-File, whereby citizens submit their tax returns and payments through commercial firms. As part of this service, companies such as H & R Block collect citizen tax forms, and electronically transmit them to federal and state governments. By 2001, the number of people in some states relying on this option dwarfed all others combined. For example, in Illinois, 957,297 people filed through E-File, compared to 147,014 using TeleFile, 62,556 employing Internet filing, and 108,501 using personal computer software such as TurboTax.

NetFile (or I-File) represents the most recent (and the most truly electronic) option. Under this system, which is available in many states (but not the federal government due to opposition from commercial tax preparers), citizens file their tax returns online for free through the state's Internet web site. This allows taxpayers to bypass commercial tax preparers and file returns directly with their state's revenue department. While the number of people using this method is lower than other options, its growth rate is expanding. In Colorado, for example, 30,360 used it in 1998, the first year it was available. By 2000, that number had grown to 111,004.

Commercial firms have opposed the expansion of Internet tax filing at the national level because they argue private companies can do so more cheaply and efficiently than the national government. Bernie McKay, vice president for software company Intuit, which sells Quicken TurboTax, successfully pushed for a federal provision that enabled private companies to control online tax filing as long as they allowed the poorest taxpayers (those making under $25,000) to file for free. In 2002, around one million taxpayers relied on that free service. Most other taxpayers having simple returns had to pay $10 to transmit their returns to the government.[4]

The George W. Bush administration initially opposed this industry request, however, saying that the federal government should offer free online tax filing similar to what exists in many states. If one seriously wanted to boost usage levels, officials argued, it had to remove any disincentive people faced to file their taxes electronically. The easier it was to use the Internet for tax filing, the more citizens would turn to electronic transmission.

By late 2002, though, after a series of extended negotiations, the administration reached a three-year agreement with seventeen commercial tax preparers that provides free tax filing for nearly three million people.[5] Using a link from the Internal Revenue Service website called Free File, low income taxpayers can go to a private site and file federal tax returns free of charge. This thereby provides a filing mechanism for the poorest segments within society.

H & R Block provides free filing for any individual with an adjusted gross income of less than $28,000. TaxACT.com meanwhile offers free-of-charge returns for anyone with an adjusted gross income of up to $100,000. Other firms provide free tax computing to military personnel on active duty, those over the age of fifty, people who qualify for the Earned Income Tax Credit for poor individuals, and those living in selected states of interest to a particular company (such as Michigan, Wisconsin, New York, Illinois, and Georgia).[6]

But this agreement with the federal government did not include state or city governments. Many of those entities already had developed their own, free, online tax service, which taxpayers directly access.[7] Cities and states had no incentive to agree to the demands of commercial tax providers that the public sector not create Internet tax filing on official government websites.

Consumer groups have complained that the federal government agreement did not protect citizens from commercial pitches. Some firms make money off of consumers by advertising "refund anticipation loans" at high rates of interest. Since the IRS requires commercial firms to pay refunds within ten days of deposit in an e-file account, these loans are quite lucrative for the private sector. Alternatively, companies such as H & R Block provide free tax filing but charge $29.95 for a "professional review" of the return. Chi Chi Wu of the National Consumer Law Center complained that companies were abusing the program through expensive add-ons. "Are they going to charge customers five bucks a pop for every question they have? What happens if you need to make a customer-service call?"[8]

Following these complaints, pressure from consumer advocacy groups in 2003 forced the IRS to make changes in this program. Most of the private companies agreed to quit providing refund anticipation loans, some of which had interest rates of up to 67 percent. All vendors furthermore were required by the IRS to provide warranties for their tax filing software. If as a result of software errors taxpayers underpaid their taxes and thereby incurred late-filing penalties and interest, the commercial vendor would be responsible for the late charges.[9]

Despite the push toward online tax filing, most Americans still rely on old-fashioned mail delivery. According to a national public opinion survey conducted by *USA Today*, CNN, and the Gallup organization, two-thirds of Americans mail their tax returns rather than file electronically. The reasons people give for this preference include: 25 percent who say they leave the decision to their tax preparer, 19 percent who do not want their financial data on the Internet, 16 percent who do not want to pay the fee associated with electronic filing, 16 percent who do not have a computer, 5 percent who say electronic filing is too much trouble, and 19 percent who cite some other reason.[10] Thus, while one-third make use of online filing, large numbers remain outside the realm of this electronic service-delivery option.

NATIONAL TAX FILING

In the United States, around 31 percent (40 million) of the 130 million individuals who report federal income tax returns filed electronically in 2001. Of these, around 32 million were delivered by tax-return professionals and 8 million were computed by taxpayers themselves. This total is up from the 11.8 million who did so in 1995. In 2002, the number of online filers jumped to 53 million (41 percent), while in 2003, 60 million (or 46 percent) filed electronically.[11]

The rate of increase from year to year has been slower than anticipated, however. Between 2000 and 2001, there was an increase of 13.7 percent, which was about half the level anticipated by the IRS and significantly below the percentage of preceding years. Between 2002 and 2003, the number filing electronically increased by 15.4 percent. This led IRS officials to question whether they will meet the national goal of 80 percent electronically filing by 2007.

In 2000, the Treasury Department created an online, e-payment system for transmitting tax payments to the federal government. Called EFTPS-Online, this electronic funds-transfer system enrolled more than 20,000 taxpayers and collected over $2.8 billion in tax revenues in its first year of operation. Over the following year, the number of users rose to over 832,000 and the revenues to $21 billion. The Treasury Department spent $9.5 million developing and rolling out this system.

As a sign of interest in online tax filing, the IRS reports there has been an upswing in visits to its agency website. From February to May 2001, for example, there were more than 13 million unique visitors to the Internal Revenue Service website. This represented a 57 percent increase over the 8.3 million unique visitors the preceding spring.[12]

In general, citizen satisfaction levels with online filing have been quite high. Seventy-five percent of those filing their returns online indicated they were "very satisfied."[13] Expanding the number of online filers beyond those who already do so is complicated, however, by lack of knowledge about how to file electronically. IRS research suggests taxpayers have problems with using its "self-select PIN program" to file returns online. In around 2 million cases, confidential information taxpayers submitted to verify their identity did not match what the agency already had on-file about the person. This inconsistency prevented taxpayers from submitting the return. Further, the absence of training programs made it difficult for the IRS to educate paper-filers about the advantages of online tax filing and showed how organizational factors often slow the spread of new technology in the public sector.[14]

Processing costs to the federal government for electronic filing remain slightly below that of paper returns. Whereas it cost $4.28 to process a mailed-in paper copy, electronically filed returns cost $4.14 to process. By 2007, the IRS projects that the latter figure will drop to around $2 per return as more and more people avail themselves of electronic filing.[15] If this happens, it will produce a major cost savings for the federal government, and free money for other budget priorities.

IRS website delivery speed—the amount of time it takes the site to come up on the screen after visitors log on—ranged between 2.5 and 3.5 seconds in spring 2001, but improved to 2 seconds after April 15, the tax-filing deadline. An analysis of availability, which is "the percentage of time the Web site's home page downloads fully," shows that during peak filing months the IRS website fully loaded only 65 to 75 percent of the time.[16]

STATE TAX FILING

The percentage of national tax filers is comparable to that at the state level, according to data provided by the IRS through its FSEF program. As shown in table 5-1, nearly 46 percent of tax filers in South Carolina file electronically, while at the low end 22.6 percent of New York residents file their returns electronically. The average rate across all states was 32.2 percent in 2002, making the state figure a little less than the percentage for the federal government.

The roughly similar nature of the state and federal figures suggests that the fact that states allow for direct and free filing online while the federal government does not has little effect on citizen usage levels. Interest group conflict over the development of federal online tax filing has altered the way the latter has developed, but has not constrained the ability of federal decision makers to offer choices through commercial tax preparers that encourage citizens to file electronically. Vociferous group lobbying has shaped the way in which online tax filing has unfolded, but not citizen usage levels.

TABLE 5-1
Percentage of State Income Tax Payers Who Filed Electronic Returns, 2002

South Carolina	45.7	Oklahoma	34.3	Washington	29.7
Mississippi	42.2	Wisconsin	33.3	Arizona	29.3
Tennessee	42.1	Nevada	33.2	Colorado	28.6
Georgia	41.2	Florida	33.2	Maryland	27.5
Iowa	41.1	Nebraska	32.4	DC	27.5
Arkansas	41.0	West Virginia	32.3	Oregon	28.4
Minnesota	39.1	Montana	32.2	Pennsylvania	27.4
Alabama	38.4	Delaware	32.1	Alaska	26.4
Kentucky	37.6	North Dakota	31.7	Connecticut	26.1
Louisiana	37.6	Ohio	31.5	Maine	26.0
Indiana	37.5	Michigan	31.3	Massachusetts	25.7
North Carolina	36.6	Illinois	31.2	Hawaii	24.7
Missouri	36.3	South Dakota	31.2	Rhode Island	24.2
New Mexico	36.1	Idaho	31.1	California	24.0
Texas	35.0	New Hampshire	31.1	Vermont	23.0
Kansas	34.5	Virginia	31.1	New Jersey	22.7
Wyoming	34.3	Utah	30.9	New York	22.6

Source: USA Today, "States' E-File Rank," April 15, 2002, p. 8A.

CHANGE OVER TIME IN STATE FILING

If one examines tax filing over time, there has been a major increase in the percentage of people in various states who file tax returns electronically. I collected detailed information from a number of different states around the country, and the data show a major increase in the rate of online tax filing. Every state has shown increases in online tax filing between 1998 and 2001. In most states, these numbers have doubled, and in some places, even tripled.

By 2001, anywhere from a quarter to a third of citizens in most of these states were filing tax returns online through various formats (whether E-File, NetFile, or TeleFile) (table 5-2). For example, in Illinois the figure doubled from 11 to 22 percent, while in Missouri, it rose from 11 to 32 percent.

Clearly, there have been significant improvements over a relatively short period of time in the percentage of the population relying on online tax services. As more and more people become familiar and comfortable with online tax filing, there should continue to be progress in this area. Since a number of states around the country have created ambitious benchmarks for electronic service delivery (80 percent usage over the next few years), they are providing specific incentives for citizens to rely upon online service features. These range from training and public education about online tax filing to offers of speedier returns and greater accuracy in the processing of tax returns.

Table 5-2
Percentage of Various States Filing Income Taxes Online, 1998–2001

	1998 (%)	1999 (%)	2000 (%)	2001 (%)
Arkansas	21	26	NA	NA
Colorado	16	23	28	NA
Delaware	NA	NA	NA	30
Illinois	11	13	19	22
Iowa	14	20	27	33
Louisiana	14	17	22	25
Maine	NA	14	21	NA
Maryland	NA	12	16	21
Massachusetts	6	10	13	NA
Michigan	11	14	18	30
Missouri	11	19	29	32
Montana	10	13	17	24
New Mexico	NA	NA	25	NA
Pennsylvania	NA	1	2	3
South Carolina	NA	NA	39	NA
Vermont	NA	NA	NA	12
Virginia	NA	NA	NA	20
Wisconsin	13	17	21	26

Source: Author's data collection and tabulation

COSTS AND FEES OF STATE FILING

One factor that has constrained the development of online tax filing is the cost of implementing and maintaining this service. It is not enough to look just at citizen demand for online tax filing and usage of service options, but also at how much it costs the public sector to put together this kind of capability. There is no doubt that financial resources are a major factor in facilitating the spread of online tax filing, nor that limited resources in some jurisdictions have prevented those areas from developing electronic tax filing options.

This demonstrates the role of budget scarcity in limiting the diffusion of technological innovation across the various levels of the American government. States that are poor have been slow to develop online tax filing; they have not managed to assemble a political coalition that can secure the financial resources necessary for this electronic service. Electronic tax filing is an expensive undertaking. To see how much it costs and the degree to which state governments have relied on in-house versus private contractors, I collected budget information from each state about online tax filing. In a lot of states, it has been difficult to develop precise cost estimates of online filing because electronic filing is not a single budget item. From interviews with agency officials, even they are not entirely clear on how many staff work on this kind of service delivery and what percentage of staff time is devoted to electronic as opposed to more traditional types of public access. Based on personal and telephone interviews with those that compiled detailed cost information, however, we were able to develop estimates on how much it costs to build the infrastructure for this kind of public service.

Most states have relied on outside contractors for the development of their online tax filing. As shown in table 5-3, nine of the 16 states (56 percent) that provided detailed information on this subject outsourced the contract to private vendors, 3 (19 percent) performed the duties in-house, and 4 (25 percent) used a combination of outsourcing and in-house work (sometimes after hiring an outside vendor, being dissatisfied with the work product, and shifting to an in-house workforce). The commercial companies involved in developing these systems include GovConnect of Cincinnati, Ohio, National Information Consortium (NIC) of Overland Park, Kansas, General Electric Information Systems, and a variety of local businesses.

Virginia had the most expensive electronic tax service of all the states. Its program cost $10.6 million over a five-year period running from 1998 to 2003. This included $947,860 in year one, $1,743,702 in year two, $2,503,266 in year three, $2,670,091 in year four, and $2,728,983 in year five.

Massachusetts also ran one of the more expensive online tax filing systems in the country. Between 1995 and 2001, the state spent $4,531,992 on its E-File program. This included $444,389 in 1995, $733,539 in 1996, $401,761 in 1997, $567,120 in 1998, $500,395 in 1999, $1,014,814 in 2000, $869,974 in 2001. This included expenditures on telephones, software, consultants, hardware, and maintenance, according to a February 11, 2002, state memo.

Louisiana had an expensive program that cost around $1.8 million to operate. This included $377,361 in personal services, $1,000,000 in hardware, $270,000 in professional services, $80,000 in software, and $80,000 in other expenses.

In looking at cost figures for online processing, development costs in other states varied from a low of $20,000 in New Mexico and $25,000 in Iowa for web-based systems to $200,000 in South Carolina and $350,000 in Maryland. Annual maintenance and operating costs once a system was put in place ranged from $5,000 in Wisconsin to $520,000 in South Carolina.

There is substantial variation across states in how sophisticated their electronic filing options are, how complex their tax code is, and what kind of software and hardware they employ to support online tax filing. The direct cost comparisons do not always mean one state is spending money more efficiently than others. Sometimes, the differences shown are due to qualitative variations across the states. But the figures allow for a comparison of the total effort being devoted to online tax filing at the state level.

TABLE 5-3
Development of Online Tax Filing in Various States

State and Website type	Developed In-House or through Out-Sourcing	Average Cost Per Electronic Return	Total Annual Cost
Arkansas	Outsource to Jim Hobson/Mountain EDI	NA	NA
Colorado	Outsource to Enablx	$0.42	$241,384 (2002)
Indiana (E-File)	Outsource to Six Companies	$0.10	NA
Indiana (Net File)	National Information Consortium	$1.00	NA
Illinois (TeleFile)	In-House	$1.17	NA
Illinois (Net File)	In-House	NA	NA
Iowa (FSEF)	Federal Government	$0.20	NA
Iowa (TeleFile)	GovConnect	$4.10	$25,000 (1999 development) $71,500 (2001)
Iowa (WebFile)	GovConnect	$50.00	$200,000 (2000 development) $51,500 (2001 maintenance)
Louisiana	In-House	$0.05	$377,361 (2001 maintenance) $50,000 (2001 development)
Maine	Outsource to GovConnect for Telefile; In-House for E-File and Net File	$1.51	$187,200 (2000 development) $178,992 (2000 maintenance)

Maryland (E-File)	GE Info Systems	$0.39	$214,604 (2001)
Maryland (Net File)	In-House and GE Info Systems	NA	$350,000 (2001 development) $35,000 (2001 maintenance)
Massachusetts (WebFile)	In-House and Outsource to GovConnect	NA	$180,000 (Staff) $869,974 (Hardware/Phones)
Missouri	In-House	NA	$245,000
New Mexico	In-House and Outsource to Gen Service, Inc.	NA	$20,000 (1997 development) $350,000 (2001 maintenance)
Pennsylvania	Outsource to GovConnect	$1.25	$390,000 (2000 development) $70,000 (2001 maintenance)
South Carolina	Outsource to Impressa and In-House	NA	$200,000 (2000 development) $520,000 (2001 maintenance)
Vermont	OutSource to GovConnect	NA	$124,000
Virginia	Outsource to American Management Systems	NA	$947,860 (1999 development) $2.5 Million (2001 maintenance)
Wisconsin	Outsource to Private Company	NA	$200,000 (2000 development) $5,000 (2001 maintenance)

Source: Author compilation

In order to more fairly draw comparisons, some states calculate cost figures on a per-electronic-return basis. This means that they total the costs of running online tax filing and then divide those figures by the number of citizens making use of the service. This allows the government agency to gauge the specific cost of each service provided and see how it might be able to achieve economies of scale as more and more people take advantage of the service. Since the expectation is better service delivery at lower cost, the hope is that as more citizens file electronically, there will be a drop in the per-return cost of tax filing.

In general, cost estimates from the various states ranged from 10¢ a return for electronic filing in Indiana to $50 a return in Iowa. Obviously, the figures vary depending on how big a user-base a state has to spread out the costs of the application, plus the expense of the contract to service electronic filing. States that have made greater progress in encouraging citizens to rely on electronic services reap greater dividends in terms of per-unit cost savings.

There are clear savings between electronic and paper tax filing. For example, Colorado reports that costs for processing electronic returns were only half as expensive as the price of paper returns. Whereas it took 42¢ to process an electronic tax submission, it cost 96¢ for each paper return that was processed. This cost differential mainly is due to the greater number of person-hours required to process paper tax returns.

Part of the efficiency gained through online filing is that the government does not need staff to enter information electronically from paper copies. The material arrives at the agency in electronic form, which saves considerable effort on the part of the revenue department. This is one of the ways the public sector economizes on electronic services and a clear reason why governments have an economic incentive to move to online services and encourage their citizenry to take advantage of available options.

Nearly every state surveyed reported reasonable satisfaction in terms of paying for the costs of implementing online tax filing. Based on our interviews, officials felt their particular payment system (either outsourcing or in-house development) had performed well, that they had a viable system, and that online filing was achieving cost savings for the particular agency.

There were growing pains, however, in a few states. In Colorado, for example, revenue staff member Valerie Horwitz reported that the initial costs of her state's NetFile system was a "source of angst." According to her, "a prior director (and prior administration of the state in general) tried to innovate, but used hourly contractors to supplement our IT shop and we were badly burned. All that remains of that system is the individual income tax return to the Internet, which has been replicated by a private sector provider, Enablx."[17]

Once a system is put in place, most states appear able to operate online tax filing with a few staff members (two to six IT specialists generally between earning $30,000 and $40,000). This is a relatively inexpensive system to operate given the large numbers of citizens who can be served and the convenience that is provided. Even these systems, however, require maintenance and periodic updates as the tax code changes over the course of time.

For states relying on the IRS to forward state tax payments and forms, costs are even cheaper. The expense to the state of maintaining this system is minimal since the federal government bears the bulk of the processing expenses. The IRS provides the staff and labor required to check returns and tabulate results, which means the cost to the state is relatively modest.

Based on these cost figures, it appears that the expense of online tax filing is reasonably low and there are clear economies of scale that can be achieved. Once such systems are put in place and 10 to 15 percent of the general public files returns electronically, the potential for cost savings and administrative efficiencies rises dramatically. In short, online tax filing is a powerful example of successful technological change that is pleasing to government officials and the general public that uses the service. Despite these advantages, however, it is not transforming the public sector because officials have succeeded in convincing only about one-third of current taxpayers to make use of this online service. Until larger numbers of citizens take advantage of Internet services, e-government's transformational potential will be limited.

Success of Technology

In terms of public satisfaction with online tax filing systems, most states reported extensive success in implementing the electronic service. Based on comments received from the public and reactions from officials who run the tax programs in various states, the technology appeared to function effectively and taxpayers are happy with the results. Other than routine, scheduled maintenance, the online tax filing system does not go down and is available twenty-four hours a day, seven days a week, as advertised to the general public. States report having an infrastructure whose capability exceeds the number of people wanting to file tax returns online, at least when based on current usage levels.

Even during peak time, the first two weeks of April before taxes are due on the 15th, there have been few citizen complaints regarding the accessibility of online tax filing sites. There was more of a problem reported with TeleFile than NetFile. For the state of Illinois, for example, only 64 percent (147,081) of the total calls (229,722) to access TeleFile were successful. Few complaints, however, were recorded about NetFile, and most citizens said they found the system easy to use.

In Maryland, based on state memos, the General Electric network that supports online filing is "configured to achieve an internal target network availability of 99.8 percent with an internal goal of 99.95. Warranted system up time is 97 percent. Our internal service-availability level has consistently exceeded 99 percent." According to web evaluations reported on the agency's website, 1,739 out of 1,805 (or 96 percent) rated the online tax filing site as excellent, very good, or good. Sixty-four percent of the total rated it excellent overall.

Pennsylvania reported some problems in terms of service interruptions. A database problem caused multiple service slowdowns, although service never was completely lost. This led to some citizen complaints and times when the service was unavailable to those wanting to file taxes.

Vermont, too, noted some problems. For example, a Freedom of Information Act request concerning citizen complaints yielded the following "Help" emails. One taxpayer wrote, "I have tried at least ten times to file electronically . . . I wish you had kept TeleFile in place, at least it was user-friendly." Another wrote, "I have tried TWICE on different days to use the V-File system to no avail. I input the data, click the button to SUBMIT, and receive an error message . . . this is supremely frustrating." One user who described a similar problem received a reply from the Vermont Department of Revenue saying, "we're in the midst of looking into this problem, and we believe we have a cure."[19]

South Carolina, however, reported generally positive feedback to its online system. Citizens sent in the following comments to the state's revenue department about its electronic tax-filing website:

> I'm sorry to have to use this format, but you have no comments section. I just completed my taxes online using SCnetFile.com and it was so easy. Thanks to whomever put this thing together! It was quick and painless. Someone did an outstanding job. Just thought you'd like to know someone appreciates the hard work.

> My hats [sic] off to you! I just finished doing SCnetfile for the 1st time. What a great thing. I wish you could help the IRS do the same thing. It is frustrating as a tax payer with a simple return to have to buy new software each year and to have to pay to submit my taxes electronically. It's 2002 for goodness sakes.

> I just completed filing our SC tax return information for 2001 using SCDOR's e-file. What a way to file! Absolutely the best thing since apple pie, chocolate cake, and vacations in Hawaii. A big THANK YOU to the SCDOR!!!!

> I thought that the filing went very quickly. . . . The most difficult part was figuring out how to switch off of AOL to the other browser. I really don't think that it could have been any easier!

> I received my tax refund this past Friday. The SCDOR should show the IRS how to set up their e-file program for on line e-filing. I am really impressed with your sys-

tem. Good work and thanks to everyone who helped set up this marvelous avenue to filing.

I just finished using the SCNET internet service. I want to tell you that the software was one of the most easiest I have ever used.

I would like to mention that I did have trouble with putting a capital gain distribution from my mutual fund (line 13 of Federal Schedule D) without first using a false "$0.00" entry in long term stock sale "fill in" section of your software. It took several tries to figure out how to get past this so that I could make the long term capital gain entry in the Section III of your developed Schedule D form.

My manual computations determining my refund and your program were the same.

Many Kudos to the software developer. I look forward to using the software next year.

Virginia collects data on citizen satisfaction with online tax filing through its website. Overall, 97.9 percent of NetFile survey respondents said they found the service easy to use. Ninety-nine percent said they would use it in the future. When asked what the major benefits were of using the online tax service, 45 percent said "faster and easier to use than a paper return," 28 percent named "confirmation that the return was accepted," 16 percent cited "email verification when my refund was processed," and 10 percent indicated "email reminders of information regarding the status of my filing."[20]

To summarize, of those people using online filing, there is high satisfaction with the service. Citizens generally find electronic filing easy to use. They enjoy the convenience and time savings associated with online services, and the quick refunds that they get through filing their returns online. Most taxpayers reported few specific problems and the systems in most states were fully adequate to meet the usage levels that currently exist.

THE LIMITS ON TECHNOLOGY DIFFUSION

Government officials have made good progress on developing online tax filing, but usage levels reveal that two-thirds of Americans remain outside the realm of this online service. Usage follows a model of incremental change rather than large-scale transformation. Satisfaction among customers is high but growth rates have been more modest than projected by state and national tax agencies. There are a variety of organizational, political, and financial reasons why citizen usage is not higher.

The biggest challenge facing the public sector has been putting systems in place that take full advantage of the power of the available technology. At the national level, an agreement between the Treasury Department and a consortium of commercial tax providers aims to boost online tax filing without the federal government developing its own infrastructure. Through links from the IRS website as well as FirstGov.gov, private companies will provide free, online tax filing for up to 60 percent of American taxpayers—around 78 million people altogether. Rather than having to bear the cost of creating its own online tax system, the federal government will rely on private tax preparers to transmit returns to the IRS.

Current usage of Free File remains limited, however, and there are few signs usage will increase much in the future. For example, a 2004 survey of Internet users who have not visited the Free File site found that 9 percent said they were very likely to visit the site the following year, 23 percent were somewhat likely, 19 percent were somewhat unlikely, and 49 percent were very unlikely to visit the site.[21] The conflict between federal government administrators who wanted to develop online tax filing and commercial providers who did not want the IRS developing this capacity is a clear example of political factors limiting the diffusion of technology. Without intense lobbying by private interest groups, the IRS was poised to move in the same direction that many state governments have gone: developing its own infrastructure for direct citizen filing of tax returns. It was only when these groups objected that the IRS stopped its development and signed an agreement allowing taxpayers to file returns through commercial preparers.

At the state level, many agencies have been hamstrung by a lack of online mechanisms for credit card payments, debit card withdrawals, or electronic fund transfers. Illinois, for example, has implemented credit card payments and in the first year, there were over four thousand credit card transactions. The state, however, would like to add a direct debit card program in the future. Missouri would like to implement an Electronic Funds Withdrawal system for taxpayers. This would allow the state to directly withdraw funds from the taxpayer's checking account in order to pay tax liabilities.

Nearly every state reported plans for future enhancements to their online tax filing systems. While a number of states have web-based systems for online income tax filing, most would like to put online comparable systems for paying corporate, sales, and specialty tax obligations. Most focused on individual income tax systems first because of the large number of such payers and the costs of processing paper returns.

Another thing states would like to do is be more proactive about public education activities. Rather than waiting for citizens to arrive at their websites and use online services, some states would like to go out and educate citizens about the availability and usefulness of the service. This would promote online filing as an alternative to paper filing and help citizens learn to become more comfortable with electronic options for state service delivery.

In order to chart their own progress, a number of states have established benchmarks for the percentage of citizenry using online tax filing. Iowa, for example has established a goal of having 80 percent of income tax filers using the online service by 2007 (the same as the federal government); Maryland has the same percentage goal, but by 2004. In order to accomplish this, Iowa is "planning a cooperative effort with local libraries around the state to encourage the use of the town's library as a resource center from which the Internet can be used to file returns electronically." Public officials also are making speeches to the effect, "Get your refund in about 2 weeks! E-File your 2001 Iowa income tax return! Don't file by paper—you'll wait 12 weeks for your refund!"[22]

Through these and other types of public outreach efforts, governments hope to make citizens more at ease with online tax filing. Having created the technology that enables people to file electronically, the challenge now is getting more and more people to use this service. Even in states that have developed sophisticated applications, usage levels remain around one-third of taxpayers. Of states surveyed in 2001, Iowa had the highest participation at 33 percent, with others falling below that level. At the national level in 2001, 31 percent reportedly made use of online tax filing.

This means that most Americans still fall outside the realm of online tax filing. Both state and federal governments have optimistic plans to boost participation by 2007, but it is unclear whether these targets will be met. Part of the challenge with any new technology is making people comfortable with the invention; the United States has a "digital divide," meaning many Americans do not have computers and do not use the Internet. Public opinion surveys suggest that nearly one-third of Americans make no use of the Internet.

If online tax filing levels continue to lag state and federal projections, it will limit the cost savings that are generated via this particular electronic service. Large efficiencies occur only when substantial numbers of taxpayers make use of the electronic service. Governments have tried different types of incentives for online filing, but it may take direct financial incentives to boost electronic filing significantly. Revenue collection agencies could offer small cash rebates such as $50 for first time, online tax filers. This would give people a material incentive beyond faster refunds and time extensions to encourage them to switch to online filing.

Some citizens have held back from accessing electronic services for fear over whether government agencies do an adequate job protecting the privacy and security of online transactions. The number of "hacker" attacks into government websites concerns Internet users. People must feel comfortable that filing taxes electronically and completing other kinds of online services will not compromise their personal information.[23] Unless government websites reassure visitors (something that is becoming more common but still remains a major challenge), it will be hard to spread e-government usage beyond the educated and affluent who already rely on them.

Until usage levels rise substantially, it will be a major challenge for government officials to use technology to spearhead a transformation in public sector service delivery.[24] Citizens must actually make use of online services to achieve the economies of scale that will lower the per-unit cost of providing these kinds of services. Barring increases in public usage, service delivery will remain the purview of the elite and more well-to-do, and governments will not realize the cost savings hoped for from online service delivery. In this regard, there is a direct connection between lowering the digital divide in citizen usage patterns and budget savings. The Internet will not transform the fabric of the public sector until more users access electronic services.

CHAPTER 6

Public Outreach and Responsiveness

TECHNOLOGY ADVOCATES often have touted the potential of new inventions to transform civic life and bring citizens closer to government.[1] At the time they were introduced, for example, the telegraph and telephone were considered major opportunities for improving mass communications. With each device, citizens could transmit information much more rapidly. This improvement in "real time" cross-continental transmission made it easier for people to find out what was happening all around the country.

Both inventions proved to have tremendous value to news organizations in their efforts to cover far-flung events. Starting with the Mexican-American War in 1846, the telegraph allowed newspapers to transmit information across broad geographic areas. Reporters operating at the front line could relay news about daily battles for publication the next day. While it took telephones a longer time to diffuse among the general population, they, too, were very helpful in facilitating communication. People could call friends and family, and keep in closer touch with what was happening in their lives.

Similar arguments about improvements in civic life resulted from the invention of the radio. When this device came into being, it brought an instancy and immediacy that was lacking in earlier communications technologies. Radio stations could provide hourly updates on the news and broadcast live political events to people who were geographically distant from the event. These features made radio the medium of choice during the 1920s, 1930s, and 1940s, when the Great Depression and World War II generated extensive public interest. Radio gave people direct and up-to-date information that was very beneficial to the relationship between citizens and leaders. Citizens felt closer to events of the day, and this was important for the relationship between the general public and elected leadership.

More recently, inventions such as two-way cable television, electronic mail, and the Internet have been cited as examples of technologies that would improve the functioning of society and government. By facilitating communications and adding an interactive component that allows people to send and receive messages, each of these devices has the potential to democratize mass communications. Rather than being stuck in a hierarchical world in which leaders send one-way messages to citizens, the interactive nature of these technologies creates the opportunity for public outreach on the part of government and better responsiveness by public officials to the wishes of ordinary folks.

Responsiveness and public outreach are desirable from the standpoint of democratic governance because they are central to representation. There are many different versions of democratic thinking, from representative to direct democracy. But every model emphasizes communications, citizen involvement, and leadership responsiveness. Regardless of whether citizens exercise influence through participation and decision making or indirectly through leader selection and democratic elections, it is important that leaders and citizens communicate with one another and have information that improves accountability and representation.

Robert Dahl is an example of a democratic thinker who emphasized representative democracy.[2] In his vision, citizens did not make public policy directly but rather exercised control through their choice of leaders. Competitive elections between different slates of candidates were key to the functioning of democracy. Citizens require clear communication during campaigns and need leaders who are responsive to citizen preferences.[3] Without these ingredients, political systems are not effective at linking citizens and leaders.

Advocates of direct democracy represent a very different point of view. Rather than counting on political elites to represent citizen interests, writers such as Carole Pateman suggest citizens should play a direct role in decision making.[4] This can take the form of town meetings at the local level, policy referenda at the state level, or public opinion polling at the national level. In a situation where citizens make policy themselves, it is even more crucial that they have good information and that there are close ties between their policy wishes and governmental decisions. A political system in which citizens participate directly in the crafting of policy requires quality input from the public and mechanisms for the transmittal of public preferences to the collectivity.

In this chapter, I look at digital technology from the standpoint of public outreach and citizen responsiveness. Using the case of e-government, I examine how much new technology has been integrated into government agency websites. Do these sites reach out to citizens and use the interactive features of the Internet to offer helpful features? To what extent are governments incorporating electronic attributes that enhance responsiveness and accountability on their websites? Briefly, I find evidence that is consistent with models of incremental change. Most public sector sites have not made much progress at incorporating public outreach or personalization into their websites. And on the crucial dimension of responsiveness, email and the Internet have not enhanced leadership responses to ordinary citizens. Through an email responsiveness test, I show that the level of responsiveness to public information requests has declined for most websites. This demonstrates that government officials are not using new technology to alter the relationship between leaders and citizens.

E-Government and Democratic Representation

E-government planners have emphasized the ability of digital technology to transform the public sector by bringing citizens closer to government. By its very nature, this communications vehicle is convenient and easy to access. Its interactive features are designed to facilitate communications between people and leaders. Rather than having to call a government agency, mail a document, or visit a public office, citizens can access information and services online, regardless of when the agency is open and without the need for special political connections.

In its ability to facilitate communications and responsiveness, digital access fits within a long line of technologies thought to improve the relationship between leaders and citizens.[5] New inventions from interactive cable technology to email have stimulated hopes that government officials would become more responsive to ordinary citizens and that there would be a closer tie between officials and the general public. By improving responsiveness and public outreach, the hope has been that technology will strengthen democratic institutions and increase public support for government.

Yet while the technical capacity to facilitate greater responsiveness is available, many government sites have not taken full advantage of new possibilities. As pointed out in earlier chapters, many public sector websites do not offer online services. There remain barriers to usage for special populations, such as those who are disabled, not highly literate, or are non-native speakers.

In addition, government officials have not made much of an effort to incorporate democracy-enhancing features onto their websites. There is little emphasis on accountability or representation on public sector sites. Instead, government officials bring a service-delivery mentality as opposed to a vision through which the Internet could transform the political system and improve the functioning of democracy.

The Technology Vision

Ever since societies have grown too large for town meetings to be feasible, direct democracy has been nearly impossible to implement. The logistical problems of bringing large groups of people together have hampered democratic institutions, as has the heterogeneity of interests within society. As society has increased in size and moved through industrial and postindustrial stages of social and economic organization, these barriers have become too large to be overcome. Instead, societies have been forced to rely on representative democracy, whereby elected and appointed leaders act on behalf of ordinary citizens. At the local level, mayors and city council representatives pass laws that apply to everyone else, while nationally, members of the legislative and executive branches enact bills and statutes. If ordinary citizens do not like the actions these leaders take on their behalf, their main option is to defeat them in the next election and replace them with a new representative who will act on their behalf and more closely represent their interests.

The age-old problems of this type of indirect democracy are lack of communications between leaders and citizens and questions concerning the extent to which leaders respond to constituents. Leaders do not always stay in touch with the grass roots and remain active in representing the public. They have their own opinions, they may listen to small groups as opposed to the general citizenry, or they may weigh the views of donors or friends more heavily than those of the general public.

One of the virtues of technology is that it long has been thought to be an ideal way of restoring direct democracy to large-scale societies. In an era where it is physically impossible to bring together all citizens under one roof and have them participate in communal decisions, technology offers the prospect of electronic communication and participation in community decision making. It overcomes the problem of geographic distance and disparity of interests in the representation of citizen viewpoints.

In the 1970s, for example, two-way cable systems were thought to be a major step forward in terms of connecting the general public to political decision making. Rather than merely being the passive recipient of cable signals, some communities were wired for two-way communications. Citizens could send as well as receive messages through their cable boxes. This offered intriguing possibilities for public participation in government decision making.

In 1977, Warner-Amex embarked on a widely-publicized experiment in grass-roots democracy in the town of Columbus, Ohio, through an interactive cable system known as QUBE. Later, the system was franchised in Dallas, Cincinnati, Pittsburgh, Milwaukee, Houston, and St. Louis. Under this technology, thirty thousand citizens in Columbus who subscribed to the service for around $10 a month could use their cable television connections to register preferences, cast ballots, order movies, or play games.

At times when there was a city council meeting or congressional hearing, viewers could watch the live debate and voice their own preferences regarding how the particular issue should be handled. Their cable box was outfitted with five buttons, which could be programmed to offer various choices. Buttons could be coded for yes, no, or undecided preferences. Or there could be alternative scenarios, with viewers asked to choose a particular option.

One experiment involved U.S. House Representative Tim Wirth (D-Colorado), who participated in a cable broadcast dealing with whether the federal government "should do more to support math and science education and research and development?" Following his speech, that question appeared on viewers' television screens and through the keypads attached to their cable box, viewers could vote yes or no on this funding proposition. According to news accounts, 75 percent voted in favor of doing more to support math and science education, which was the stance taken by Congressman Wirth.[6]

It was not just matters of public policy on which opinions could be expressed. Thousands of fans who watched a semiprofessional football game on July 12, 1980, between the Racine Gladiators and Columbus Metros were given the opportunity to call plays throughout the game. Depending on field position, viewers could choose one of five different plays that appeared on the screen. With an almost instantaneous tabulation of the fan preferences, the Columbus team agreed to call the play that had been requested by their fans, using an electronic transmitter in the helmet of the quarterback.[7]

The same concept was used later that year in a television comedy show, *Lulu Smith*. One episode, which involved a young woman confronting a series of life options, allowed Columbus-area viewers to decide which plot the episode should follow. Using their keypads, people could select one of five different plots for the young woman as she thought about her life. She then acted out the option that television-watchers had chosen for her in what producers called "the nation's first major 'interactive' television drama."[8]

In the abstract, these possibilities represent a fascinating illustration of how technology can improve communications and provide a mechanism for citizens to participate in various kinds of decisions. The QUBE system allowed geographically disparate individuals a chance to watch the same debate (or athletic contest or drama), and participate in the outcome of that event. People could express opinions, convey their preferences, and otherwise shape the community response to a particular set of choices.

In reality, however, the QUBE technology never fully delivered on its reputed promise. In most communities, only a small proportion of citizens actually used the service, even when they subscribed to it.[9] At $10 a month, it was considered expensive, and many citizens did not want to spend the extra money subscribing to it or taking the time to participate in collective decision making. Those who did subscribe were not representative of those who did not. Users tended to be more upscale and better educated than nonusers, which made the interactive technology little better at solving the problems of representative democracy than existing social and political institutions.[10]

Similar problems have developed in regard to other new technologies. For example, email was an invention with the potential to reshape social realities. By allowing two or more people to communicate in nonsynchronous time, email raised the possibility that communications problems could be reduced. People could convey information instantly over wide geographic areas. They could send and receive messages as quickly as they could type the words.

Yet email has generated its own social and political pathologies. The technology now generates more unwanted spam than desired communications.[11] Weight loss and sweepstakes companies send mail as do those interested in selling get-rich-quick schemes and pornography. Rather than empowering ordinary people, as appeared possible in the early days of its existence, it has become an electronic tool that annoys and disgusts users.[12]

In the political realm, there are problems related to the sheer volume of electronic messages. Members of Congress receive so many communications that they cannot respond to all of them individually. Instead, it has become common for elected officials to have auto-reply features that require email senders to reveal their postal address before the member answers. Only those people who live within the district of the particular representative get a "real" response. This is the way congressional offices deal with the flood of email that enters their in-boxes each day.[13]

More recently, there have been interactive features, such as "push" technology and website personalization, that offer new possibilities for technological change. Push technology refers to communications devices that send out material automatically to people based on specific preferences. Rather than waiting for visitors to come to a government agency and request information, push technology delivers emails, newsletters, subscriptions, or direct mail to users interested in those areas.

For example, soybean farmers could benefit from having up-to-date information on their industry. If the Department of Agriculture allowed users to register particular kinds of interests, government officials could employ push technology to send material on the latest research and market conditions to those people. Rather than being reactive, administrators could become more proactive, and help constituents learn about things vital to their interests.

Website personalization allows visitors to tailor a particular website to their own interests. Using registration of visitor preferences, websites can automatically configure themselves to emphasize material related to those qualities. For example, if a person is interested in the environment, then the website automatically can be configured to show the environmental stances of political leaders as well as their pledges and promises in that area. Or if someone prefers to know about the weather in Chicago, a website can be altered to display conditions in that locality. Personalization is a way to tame the wealth of material available on the Internet and tailor it to someone's interests.[14] This helps to customize information flows and enables the recipient to receive information based on predetermined categories.

Of course, it remains to be seen to what extent these new technologies have been incorporated in government websites and whether they actually improve democratic performance in the public sector. Many past inventions started with bold expectations only to deliver meager results in the end. Similar to the QUBE technology and email, these creations have proven unable to deliver on stated hopes. Rather than representing examples of social and/or political transformation, the change that resulted fell more within models of incremental or secular change.

This especially has been the case in the public sector, where new technologies must confront existing bureaucracies, limited budget resources, and elected and/or appointed officials who may not be sympathetic to new technology. These realities constrain the extent of social change and limit the ability of technology to produce large-scale alterations. For these reasons, it is necessary to investigate the extent to which features of public outreach and democratic responsiveness are being incorporated into government websites and how they are altering existing political realities.

PUBLIC OUTREACH AND EMPOWERMENT

There are many aspects of Internet technology that offer hope of improving the connectedness between citizens and government agencies. These include communications mechanisms that allow citizens to register ideas and complaints with public officials, web features that publicize government actions and thereby improve accountability, and attributes that give citizens greater control over accessing and using public sector resources. Indeed, the virtue of the Internet is its two-way communication and ability to tailor online material to the needs and interests of specific visitors. These are the very features that planners have extolled in their rhetoric about an "e-government revolution."

In this project's examination of thousands of government websites, we looked at specific features within each website that would facilitate connections or interactivity between government and citizens: email addresses, areas to post comments or complaints, chat-rooms that allow visitors to converse, search features that help people find information, broadcasting of government events, push technology that brings information directly to the attention of citizens, and website personalization that allows visitors to tailor a site to their own interests.

Each of these features is important because it has the potential to reconnect citizens and leaders.[15] Obviously, particular devices facilitate connection in very different ways. Some are more interactive than others. Still others are episodic in nature, rather than providing more continuous and meaningful types of feedback or interaction.

But regardless of the specific type, each one of these technological attributes is designed to improve outreach, responsiveness, and communications. By improving communications and bringing leaders and citizens closer together, these technological innovations increase public involvement in and mastery over their environment. In the long run, interactive technologies can serve as vehicles to re-engage citizens and improve leadership responsiveness to the average citizen. Both of these are key needs at a time when many citizens feel alienated and distant from leadership, and in which political institutions are based on representation and responsiveness.

In our study of state and federal websites, most of the results fall squarely within an incrementalist perspective. The most basic kind of interactivity is email, whereby an ordinary citizen can contact a person in a particular department other than the webmaster. In 2003, 91 percent of state and federal websites had email addresses, up from 81 percent in 2002, 84 percent in 2001, and 68 percent in 2000. At a rudimentary level, most sites have basic contact information, and this kind of feedback mechanism has improved over time.

Other methods that government websites employ to facilitate democratic conversation include areas to post comments (other than through email) and the use of message boards, surveys, and chat-rooms. Websites using these features allow citizens and department members alike to read and respond to others' comments regarding issues facing the department. In 2003, 24 percent of websites offered this feature, more than double the 10 percent from 2002 and nearly five times greater than the 5 percent in 2001. This shows the steady pace at which many interactive and feedback features are being incorporated into government websites.

This project also examined the extent to which government websites were searchable. In 2002, 43 percent of the sites studied had a search feature. This is helpful to citizens because it allows them to find the specific information they want, as opposed to the material government officials want to present to them. It breaks up the hierarchy of typical communications flows, and puts more power regarding the search for information in the hands of website visitors.

Another accountability-enhancing feature of government websites is the broadcasting of speeches, debates, and hearings. Leadership debates and presentations convey ideas and proposals to the public and represent a way for citizens to determine how leaders are governing and what platforms they are preparing for future action. As such, they are essential to citizen representation and crucial for citizen efforts to hold leaders accountable.

Legislative hearings represent the same type of opportunity. Hearings are a way for one set of leaders to serve as a check on others. They typically are organized for legislators to probe executive conduct or the way in which government agencies are performing their mission. For those interested in public policy, hearings are a goldmine of relevant information.

Yet despite the widespread availability of broadcasting technology and the importance of speeches, debates, and hearings to democratic governance, only 4 percent of sites in 2002 offered any kind of broadcasts. This includes presentations such as important speeches or live coverage of Senate or House of Representatives hearings and broadcasts of a governor's State of the State Address, to weekly Internet radio shows featuring various department officials.

The relative paucity of these types of presentations on government websites means Internet technology that is readily available is not being used to enhance government accountability or improve citizen representation. As such, public officials are missing a chance to increase the performance of democratic institutions.

In contrast to the state and federal sites noted above, city government websites showed more improvement in public outreach over the last three years. Seventy-one percent of websites in 2003 allowed for email contact (about the same as the 74 percent in 2002 and 69 percent in 2001) and 69 percent in 2002 provided a search engine for the website (up from 54 percent the previous year). Thirty-five percent provided an area to post comments (the same as the 36 percent in 2002 but up from the 17 percent in 2001), and 9 percent in 2002 had broadcasts of important city speeches or events (up from 2 percent in 2001).[16] This represents an improvement over state and federal sites, but all government sites fall short of what could be done had elements of interactive democracy been incorporated to a greater extent.

In short, many simple examples of public outreach have not been incorporated in government websites and are not being utilized to their full extent. Rather, the public sector has been slow to embrace these technical creations either because they are expensive to implement, there are organizational obstacles to adopting the change, or leaders do not see these electronic features as vital to the mission of their particular government agency. This reticence clearly limits the opportunity of government planners to use the Internet as a tool to transform the public sector.

PUSH TECHNOLOGY AND WEBSITE PERSONALIZATION

There is even less evidence that government websites are incorporating interactive features, such as push technology or website personalization, into the public sector. In our study of state and federal government websites, for example, only 12 percent of websites in 2003 (similar to the 5 percent in 2002, 13 percent in 2001, and 5 percent in 2000) allowed citizens to register to receive updates regarding specific issues through so-called push technology, that is, automated software that directs emails to people with particular interests. Similar results were obtained in our 2003 study of city e-government. For those sites, only 8 percent of sites offered push updates. This was similar to the 13 percent in 2002, but up from the 2 percent in 2001. With this feature, a web visitor can input their email address, street address, or telephone number to receive information about a particular subject as new information becomes available. This material can be in the form of a monthly e-newsletter highlighting an attorney general's recent opinions or alerts notifying citizens whenever a particular portion of the website is updated. There are a wide variety of ways the government can incorporate push technology and website personalization onto their sites, but most agencies are not utilizing this new technology in any form.

Even though government serves a variety of specialized interests that potentially could benefit from these kinds of outreach activities, there is little evidence they are being included in public sector websites. Interests from health care professionals to education reformers must go to a government agency website to find particular information that is of interest to them. Individuals needing welfare or health care coverage do not receive specialized information and/or service updates that could empower them and put them in a stronger position to obtain needed help.

This result is even more dramatic in regard to website personalization. In 2003, only 2 percent of state and federal sites and 4 percent of city government websites allowed citizens to personalize an agency's site to their own specific interests. These were about the same as the percentage in previous years (2 percent in 2002, 1 percent in 2001, and 0 percent in 2000 for the state and federal study, and 3 percent in 2002 and 0 percent in 2001 for the city research). This means that if someone has special interests in a particular area, there are no mechanisms to tailor public sector websites to those interests. Rather, citizens have to wade through a variety of information that may be completely outside their area of interest in order to find the small portion of the website that appeals to them.

Some state portal pages are beginning to apply this technology. For example, California and Michigan allow users to customize their sites to highlight the information that they indicate as the most important and useful. California has a "myCalifornia" feature that allows visitors to register under any of the following categories: state resident, business person, media representative, state employee, student, or tourist. The website then sends them customized information and state service links based on those kinds of interests.

In many cases, interactive features have not been adopted because they involve a completely different role orientation on the part of the public sector. Rather than merely reacting to citizens who come to them with particular concerns or problems, push technology and website personalization require officials to provide information on a proactive basis. This perspective is not very common among government bureaucrats. Indeed, most of them feel more comfortable waiting for constituents to show up at their office or on their telephone, and then attempting to address a concern.

The relative paucity of these types of democracy-enhancing features is problematic because it limits the ability of government websites to take advantage of available technology. Part of the difficulty is the question of vision and role orientation. Many public officials are more interested in using the Web as a tool for service delivery than democratic enhancement. They see the Internet as a technocratic tool for incremental change; they are less keen on employing this technology for transformational change. Due to these kinds of limits to their imagination, they are not able to marshal the political will and the financial resources to overcome the natural inertia of government agencies and their affiliates. As with any new technology, this limits the ability of government officials to figure out how to harness technology to best effect and incorporate new features into their websites.

Of course, some observers have suggested that website personalization is not a desirable feature of electronic governance. For example, Cass Sunstein complains about media sites that engage in narrow-casting appeals, saying that they encourage political self-segregation that is not desirable from the standpoint of the broader collectivity.[17] Rather than being exposed to many different ideas, citizens can sort themselves out into various political categories and surround themselves only with like-minded people. In the long run, Sunstein notes, this undermines the country's diversity of ideas and weakens people's general understanding of other individuals' perspectives.

While this criticism holds some merit in regard to patterns of media consumption, it is less relevant for electronic governance. Here, the goal is allowing people to tailor agency links and service delivery to topics that are relevant to themselves and that further their ability to gain benefits from the public sector. Tailoring government service delivery to what particular citizens need yields little risk of ideological segregation. Rather than harming citizens, this technology empowers ordinary people.

DEMOCRATIC RESPONSIVENESS

The last aspect of public outreach that was investigated concerns responsiveness to requests from ordinary citizens. How seriously do government officials take electronic requests for information from citizens? Do they respond to them and if so, how long do their responses take to arrive? What does their general level of responsiveness tell us about the ability of technology to transform government performance?

Responsiveness is a key requirement of democratic political systems. One of the qualities that distinguishes democratic from nondemocratic government is the interest in and electoral need for paying attention to those whom leaders claim to represent. Because of elections, leaders must consider the views of their constituents. Even if they are in an appointed position, the expectation is that they will respond to citizen concerns and answer basic kinds of inquiries.

One of the virtues of e-government is the ability of citizens to register concerns and complaints directly to public officials. Email is a way to express opinions to the government. As computer usage rises, this is likely to become an even more prominent form of political communication with the government. To test government responsiveness, we sent email messages in 2000 requesting official office hours to four agencies in each state (the governor, state legislature, top court, and leading human services agency) plus emails to 86 federal agencies. This was designed to see if officials would respond to a request from an unknown citizen and if so, how long it would take them to answer the question. Our goal was to simulate a question that an ordinary citizen might have for a government agency, regardless of the type of agency that was involved in the request for information. Would public officials respond to simple requests for information?

We timed responses by number of business days, and found that government officials were highly responsive to emails. Of the 286 state and federal offices contacted, 91 percent answered our query, and 73 percent did so within one day. This reflects a high degree of responsiveness and suggests that at least in this regard, electronic communications are transforming citizen-leader relations and improving responsiveness to the concerns of ordinary people.

We repeated this test in 2001 with a slightly more difficult test ("How much does it cost to obtain government documents from your agency?") in order to determine how quickly government bureacracies responded to a more specific inquiry. In 2001 the response rate was 80 percent, with 52 percent responding within a single day. Again, these rates are reasonably high and suggest that new technology enables government agencies to become more responsive to citizen requests.

In 2002, however, when we repeated the original request for office hours, only 55 percent responded to the request. Of those who responded, response times were longer. Only 35 percent replied within a single day, down from 53 percent in 2001. Four percent took five days or more to respond. This is a clear deterioration in agency responsiveness to citizen inquiries.

By 2003, when we sent sample email messages to human services departments in the 50 states asking for information regarding office hours, responsiveness rose to 68 percent (with 62 percent responding within a single day, up from 35 percent in 2002). Only 2 percent took three days or more to respond. Even though email volumes are increasing in many government offices, this change in responsiveness suggests government officials are reconfiguring their offices to deal with the volume of citizen email.

To see how public sector responsiveness compared to that in the private sector, we rely on a recent 2003 Customer Respect Group study of email responsiveness from America's one hundred largest companies.[18] That research found that 70 percent of large corporations responded to email questions and 30 percent did not. Of those responding, 58 percent did so within one or two days, 6 percent took three days, and 6 percent responded within four days. This means that responsiveness in the public sector is quite similar to what is found in the private sector.

The comparability of these results suggests that at least on this dimension, the public sector is no worse than the private sector. This is an important finding for people who feel the public sector always performs more poorly than the business world. Antigovernment groups like to complain that public bureaucracies are not responsive and therefore need to be reformed. While there are grounds for complaints about the public sector, we found the private sector did no better on the criterion of responsiveness. Poor responsiveness is not just a challenge to the public sector, but rather is one the business community needs to address, as well.

Conclusion

When looking at these overall results, it is clear that much of our evidence is consistent with an incremental pace of technological change. Government officials have not made much progress in harnessing the power of the Internet to enhance representation and accountability. They have not incorporated many interactive features into agency websites. And on the crucial email responsiveness test, there has been a drop over time in the speed and rate of response to citizen requests for information. This limits the degree of technological change, and undermines the ability of the Internet to transform the public sector.

From a technical standpoint, there is no reason why government agencies have not incorporated more interactive features into their sites. The technology exists and has been well tested. Interactive technologies are common on some sites and have been used to serve customer needs. The failure to incorporate these advances is not a problem of technology as much as it represents a problem of political will and vision. Elected officials simply do not see public outreach and improving the functioning of the democracy as important as other areas of e-government. As a result, they are missing an opportunity to take full advantage of the technology that is available.

Public officials must recognize that government websites can do more than merely deliver information and services. They represent a powerful tool for transforming democracy and reaching out to the general public. They can improve the way in which the political system operates and help those who are marginalized whether on economic or political grounds become more effective at using government resources. The Internet has the power to revolutionize the relationship between citizens and leaders, but realization of this hope takes more vision and political will than have thus far been established in the public sector.

CHAPTER 7

Citizen Use of E-Government

THE GENERAL PUBLIC is a crucial factor in the dissemination of new technology. Owing to how they think about and utilize technology, individuals either facilitate or constrain change. If consumers are open to new technology or adept at integrating new inventions into their lives, they are going to be more receptive than if they harbor negative views about technological innovation. In addition, their ability to pay for new technology affects the speed with which innovative creations diffuse among the population.

In looking at the history of technological change, for example, it is clear that some inventions have spread very slowly throughout the population, while others diffused much more rapidly. While there are many reasons for this variation, how citizens feel about technology is an important determinant of the rate of diffusion. Feelings about their willingness to make use of new creations is a vital aspect of this process.

Telephones represent a case of slow dissemination. A study of United States Census data demonstrates that it took more than fifty years for half of the United States population to acquire telephones.[1] This time lag was much longer than those for inventions such as video games and personal computers, both of which had rapid incorporations into the country's social life. In the case of individual computers, it was a matter of just a few years from the original invention before large numbers of people owned these machines.

Part of the delay in the expansion of the telephone was its high cost. In late nineteenth and early twentieth centuries, telephones were quite expensive from the standpoint of ordinary folks. It would take an average family two weeks of job earnings to pay for a telephone. In 1906, for example, telephones cost $12.50 a piece. This was at a time when teachers earned around $25 a month for instructing primary and secondary students.[2] In addition, telephones required other purchases beyond the receiver itself. Phone companies charged a monthly usage fee based on how many calls and what kinds of calls (local or long distance) were placed. Telephones also relied on batteries for their electrical power, and these batteries had to be replaced quite frequently according to contemporaneous accounts.[3]

On account of the costs of purchasing a telephone and making calls, its diffusion took quite a long time. People viewed the telephone more as an emergency communications device than one suited for routine, day-to-day communications. Regular usage would have been far too expensive for average families to afford. Even after they were installed, Americans did not make extensive use of them for several decades.

As a sign of the barriers to its expansion, telephone usage during the Great Depression in the 1930s dropped from 16 to 13 percent nationally.[4] These figures are noteworthy because they show that more than fifty years after its invention, many Americans still did not use a telephone. Furthermore, when confronted with economic hardship, usage dropped among those who already had made the investment decision in this new technology.

The slow diffusion was not due just to cost, however. Telephones are an example of synchronous communication. This means two people have to be on the line at the same time for the communications link to be formed. Not surprisingly, given this situation, it was difficult for this technology to spread to the point where enough people had phones so that the creation was effective at reaching other individuals. If someone's family and friends were not accessible through a telephone, a person's own motivation for purchasing a telephone would be sharply limited. Telephones represent an example of a new technology whose benefit was dependent on other people purchasing it.

In other cases, such as television, personal computers, and video games, an individual's ability to benefit from the new technology was not dependent on other people's purchasing decisions. As long as proper programming and software were available, it made sense for a person to buy a television, computer, or video game and enjoy its respective advantages. The asynchronous nature of these devices meant that each invention was able to spread much more rapidly than the telephone and thereby become an example of technological change.

In this chapter, I use the case of e-government to see how the public feels about digital delivery systems in the public sector. In particular, I look at citizen evaluation and usage of government websites. Drawing on national public opinion surveys conducted in 2000, 2001, and 2003, I examine how people evaluate public websites, who uses e-government, and what the differences are in types of users. My goal is to see what the case of e-government tells us about citizen feelings toward new technology.

An examination of national survey data shows that the American public is generally positive about Internet technology. People like the convenience of government websites and the ability to access information 24/7. At the same time, however, citizens worry about privacy and security and two-thirds do not make much use of e-government. Public usage has risen slowly over time, consistent with a model of incremental change. E-government visitors tend to be male, younger, better educated, and high wage-earners.

As mentioned previously, some parts of the population have the financial means and training to take advantage of the Internet, while others do not, creating what has come to be known as a "digital divide."[5] There are clear links between Internet usage and income and education levels, with those of modest financial means and educational backgrounds being less likely to make use of the Internet. Until this gap is reduced, it will be difficult for public officials to make effective use of new technology, reach the economy of scale that lowers per-unit costs, and employ technology as a tool for social and political transformation. New technology cannot produce a revolution unless citizens are amenable to the technology and willing to make use of it. Given the expense of incorporating new technology into government websites, it takes substantial usage levels on the part of the general public to obtain the economies of scale that make such upgrades viable.

In the long run, how citizens feel about e-government and the degree to which they access online materials will shape how the electronic sector functions and the extent to which the Internet produces transformational, secular, or incremental change. It is not enough to investigate what is on government websites and how new technology is being employed; one must also explore whether ordinary folks are accessing e-government and making use of its information and services. The more positive the public is about electronic technology and the more they draw on it, the brighter the future will be for digital democracy.

CITIZEN ATTITUDES TOWARD NEW TECHNOLOGY

Citizens have a wide range of feelings about different kinds of technological innovation. Some creations are warmly embraced by the public because they pose little threat and offer advantages to the individual user. Consumer items, for example, generally fall within this category. Devices such as washing machines, dishwashers, and vacuum cleaners never were controversial. Their virtues were readily apparent and it was clear they would aid the lives of those who made use of them.[6]

Other technologies, however, have aroused great fear. Rather than being seen as new creations that will improve society, they have sparked concern regarding possible ethical problems and raised doubts about a "Pandora's box" of new problems being unleashed on the population.[7] This arises either because people do not understand the invention or they fear the uses to which it will be put, or perhaps that the creation actually has been developed with such uses in mind.

Nuclear fission represents an example of complex technology that has been put to both peacetime and wartime applications. As a technology, generating energy through splitting atoms is a routine process. Early advocates hoped that splitting atoms was a way to generate electrical power through a safe, clean, and cheap means. Fission thus appeared to many to be a scientific advance that would not endanger society.[8] But as nuclear bombs dropped over Hiroshima and Nagasaki, Japan, demonstrated in 1945, fission also could be employed as a weapon of mass destruction. It was not just a neutral scientific advance or a technique for generating cheap electricity, but a means to destroy entire cities. This sobering reality demonstrates that some technologies represent both the best and worst of the human spirit.[9]

Owing to this duality, people often have feared new technology on ethical or moral grounds. There is the "Frankenstein" fear that human medical or scientific advances will create a monster that will spiral out of control and endanger humanity itself.[10] Just as Hollywood mythmakers portrayed the Mary Wollstonecraft Shelley character as a threat to society, many observers worry that new technology poses challenges to our underlying social, political, and economic fabric.[11] Rather than liberating people, technical creations can take away freedom, invade privacy, or endanger cherished social objectives.

These concerns have led to regulatory restrictions on the employment of technology. A public that worries about the uses to which new inventions are put lobbies members of Congress to regulate the use of new technology or place limits on how inventions can be deployed. Society is not helpless in the face of new creations, but can impose restrictions on how technology is used and who has access to it.

Cloning represents a recent illustration of the dilemma posed by certain scientific advances.[12] This technique, which has already recreated a sheep and numerous other animals through the duplication of genetic material belonging to an existing animal has led to fears that scientists are "playing God" and creating new life forms through means other than natural acts of procreation. Critics worry that experimentation in this area will lead to people making copies of humans without adequate understanding of the risks or dangers.

These fears have led to congressional action prohibiting experiments involving human cloning and placing sharp restrictions on the cloning of animals. For example, the George W. Bush administration proposed and Congress adopted an outright ban on human cloning. Responding to citizen concerns and organized groups who feared unrestricted use of cloning techniques, the United States government chose to limit its use and application, despite possible benefits to scientific knowledge.[13]

Similar controversies have arisen in regard to food irradiation. Scientists have developed radiation techniques designed to destroy contaminants and produce healthier and longer-lasting foods. Tests so far have revealed no health hazards from the use of irradiation, but rather substantial benefits in terms of food preservation, yet some experts and members of the general public are not wholly convinced that this change in food preparation techniques is beneficial.[14]

The Food Irradiation Campaign in Europe has organized to limit use of this technique. After the European Union's Scientific Committee on Food group released a 2002 study saying that research showing irradiation associated with some forms of "colon carcinogenesis in rats" is not relevant to policy decisions in this area, the group complained bitterly. Their argument was that the committee should "rethink its position" because the campaign had data showing irradiation does not kill all bacteria and poses significant health risks to consumers.[15]

Similar concerns have been expressed about genetically modified food. Biologists have figured out ways to alter genes so as to combine the advantages of different kinds of food specimens.[16] For example, during the winter months it always has been difficult to produce tasty tomatoes that could be shipped across the country. Instead, growers have been forced to harvest tomatoes while they are green, ship them across the country, and wait for them to ripen on store shelves. This, of course, does not produce a very flavorful product.

To deal with this problem, a company known as Calgene used genetic engineering to develop a new vegetable called the "Flavor Saver Tomato" that could be harvested after it started to ripen, thereby yielding a more tasty food item for the winter months. This was considered an advance in the production of food, something akin to the development of seed hybrids in corn crops or artificial insemination in regard to cow and horse breeding.

The European public, however, has not accepted genetically modified food. The European Union, for example, has banned the importation of some kinds of genetically modified crops on grounds that the health risks are unknown and the danger of a "Frankenstein food" product is too high. The American government has placed rules and regulations on the use of genetically modified organisms. It wants to make sure that scientists do not produce new crops resistant to commonly used pesticides and insecticides.

Not surprisingly, these controversies over various new advances and technologies have filtered down to the American public and led to sharp opinions about their use. For example, public opinion surveys conducted by the Pew Research Center in Washington, D.C., show that people feel quite negatively about human cloning. When asked how they feel about "scientific experimentation on the cloning of human beings," 77 percent of Americans in 2002 indicated they opposed cloning experimentation. When asked why they were opposed, 72 percent said they felt it was morally wrong and 19 percent believed that cloning science was not yet safe enough; the balance did not know.[17]

The American public is not very convinced that food irradiation and use of genetically modified organisms provide social or economic benefits to the community as a whole. According to the food irradiation campaign, there is "long-standing public opposition" to the use of irradiation. Its work suggested that "the technology does not offer real benefits to consumers, and that it will lead to consumers being misled over the freshness and quality of the food they buy."[18]

In general, though, while they worry about certain new techniques, Americans feel positively about technology as a whole. When asked what they thought had been America's greatest achievement during the twentieth century, 41 percent named science and technology, making it by far the top item on the list. That accomplishment dwarfed other items named such as medical advances (7 percent), war and peace (7 percent), the economy (5 percent), and the development of civil rights (5 percent).[19] Each of these items was seen as meritorious, but not nearly to the extent as were American technological breakthroughs.

This survey also asked whether particular inventions represented a change for the better or a change for the worse. Of the fifteen items posed (from the Internet and Viagra to email and telemarketing), the top items named as changes for the better were email (71 percent), the Internet (69 percent), mutual funds (69 percent), and cellular phones (66 percent). In contrast, 51 percent thought telemarketing represented a change for the worse, and 49 percent cited the cloning of sheep as a major change in a negative direction.[20]

When quizzed regarding whether inventions had improved the quality of life in the United States, a national sample of adults polled by the Massachusetts Institute of Technology found that 95 percent of American adults believed technological inventions had improved the quality of life. Only 5 percent doubted that quality of life had been improved through these types of creations.[21]

In short, people are able to distinguish different types of social and technological developments and reach divergent conclusions about each of them. While they are concerned about some uses of technology, citizens generally hold positive attitudes about technology as a whole. When the public has not been supportive of particular technologies, politicians have rushed to enact restrictions on their expansion and use. In this way, unfavorable public opinion leads to regulatory actions that constrain the adoption of new technologies. This illustrates how citizen opinion can limit or facilitate the spread of technology.

CITIZEN EVALUATIONS OF E-GOVERNMENT

Due to the centrality of citizen evaluations in the distribution of new technology, it is important to look at how the public feels about specific advances. Are they generally positive or negative? What hopes and concerns do they hold about the technology? Do they actually make use of the new invention in their daily lives and how does that use vary for different types of people?

Scientists can create the most perfect invention, yet if potential consumers are fearful about that creation, its use will grow in a more limited fashion. The way in which people think about new technology and how receptive they are to these creations are prerequisites for widespread adoption. The modern history of new technology is quite clear in highlighting the importance of the views of ordinary people in its diffusion.

In regard to e-government, citizen assessments are particularly important. Electronic governance requires tax dollars to build infrastructure and political will to get diverse agencies to work together. Unlike private sector technologies that are dependent on social and economic factors for their adoption, public sector use of new inventions is inherently an institutional question mediated by political dynamics. As shown in earlier chapters, e-government is directly dependent on policy decisions for its implementation and execution. Unless the public sector is willing to fund its development and encourage citizens to make use of government websites, e-government usage will grow very slowly.

To investigate public opinion on e-government, the Council for Excellence in Government, a Washington nonprofit organization devoted to improving the public sector, undertook a national public opinion survey in August 2000. The survey was completed by the Hart/Teeter polling consortium and relied on interviews with 1,003 randomly sampled adults in the United States. This research represents the most detailed study of public attitudes toward and use of e-government that has been undertaken in this country. The questionnaire posed 79 items regarding feelings about e-government, frequency of visiting government websites, ease of use, likes and dislikes about e-government, and support for future initiatives in this area. Overall, the survey had a margin of error of plus or minus 3.5 percent.[22]

This was followed with another survey in 2001 that looked at trends over time and whether attitudes toward e-government had shifted in light of national developments. The 2001 survey was also sponsored by the Council for Excellence in Government and completed by Hart/Teeter. It was based on interviews with 961 adults nationwide during November 2001. Several questions in 2001 were repeated from the 2000 survey in order to facilitate comparability. The 2001 survey had a margin of error of plus or minus 3.5 percent.[23]

In 2003, the Council for Excellence in Government sponsored another survey focusing on public opinion toward e-government. This research, also undertaken by Hart/Teeter, included interviews with 1,023 adults across the country in February 2003. It had a margin of error of plus or minus 3.1 percent.[24]

In general, most citizens in the United States are positive about e-government. When asked in 2000 about their visits to government websites, more than two-thirds of adults across the country gave public agencies high marks for their websites. Seventy-one percent, for example, believed that the quality of government websites they had visited was excellent or good, 26 percent felt they were only fair, 2 percent rated them poor, and 1 percent were unsure. In terms of ease of use, 48 percent found it easy to get information about a particular service or agency from a government website, compared to 31 percent who felt it was hard and 21 percent who were not sure how easy or hard it was.

These are very positive numbers for electronic technology. Given people's generally positive views about technology in general and the Internet in particular, it is no big surprise that most are favorable toward e-government, which is based on digital delivery systems. The positive sentiments about the Internet and its convenience and ease of use make people sympathetic toward electronic governance and help to create general good will toward that technology.

Users like the fact that they can receive information and services from their homes and offices. Unlike bricks-and-mortar government, which citizens have to visit, mail, or call and from which they then await action, e-government provides instant results. Visitors can access publications, search databases, register complaints, and order services.[25] These features create considerable good will among the general public.

When asked in 2003 what they thought the most important positive thing that may result from e-government, 28 percent said making the government more accountable to citizens, 19 percent stressed making government more efficient and cost-effective, 18 percent named providing greater public access to information, 16 percent indicated providing better homeland security, and 13 percent said making government services more convenient, with the balance uncertain.[26]

The Internet differs from scientific advances such as cloning, food irradiation, and genetically modified organisms in that it is not considered hazardous to people's health. It poses little risk to health (other than damage to wrists and elbows related to carpal tunnel syndrome), does not raise troubling ethical issues, and is not considered a Frankenstein-like invention. Unlike some new devices, it is convenient, relatively inexpensive from the standpoint of website users, and makes the government more accessible to those who are able to afford a computer and surf the Internet.

Not all is positive in general public attitudes about e-government, however. People have deeply rooted fears about e-government with respect to its possible impact on security and privacy. When asked in 2000 to indicate the aspects of e-government that concern them the most, 48 percent cited hackers breaking into government computers, 40 percent cited government employees misusing personal information, and 35 percent feared having less personal privacy.

People's worries about security and privacy have been aroused by incidents in which public and private sector systems were compromised. Some have involved sophomoric humor, such as when hackers broke into the U.S. Justice Department site and placed a pornographic picture in place of then Attorney General Janet Reno.[27] Other attempts have proved to be more serious, however. Over a period of several years, a group known as "Moonlight Maze" has stolen thousands of Pentagon secrets.[28]

In the last decade, there has been a tremendous increase in efforts to break into computer systems. From a low of 6 security breaches in 1988, the number has grown to 52,658 in 2001, according to a research team at Carnegie Mellon University.[29] This is a startling rise in computer security problems. Therefore, it is little surprise that the public is worried about privacy and security on government websites. Empirical evidence demonstrates that these fears are well grounded.

Of course, e-government does not appear to most Americans to be nearly as socially dangerous as cloning. But the loss of privacy and confidentiality, and the use of the Internet to deliver pornography, hate, and racism has raised serious concerns among many Americans. Computer-users worry that they will not be able to maintain proper security and that spammers and hackers will use the technology for destructive ends.

Given public concerns about e-government, we looked at what factors were associated with citizen evaluations of the quality of government websites. There are several demographic and political factors that are thought to affect people's views about government.[30] Sex, age, and race are demographic qualities that affect how people evaluate governmental initiatives. Education typically is linked to technology because it provides experience with and understanding of new inventions. Party identification is linked to a wide variety of social and political attitudes so we wanted to determine how it related to e-government assessments.

To see what factors accounted for citizens' feelings, we undertook an assessment of how these factors linked to evaluations about the quality of government websites. In general, there is variation in how positive various types of people are. As shown in table 7-1, a significant difference exists in how visitors rate the quality of government websites. For example, Democrats were more likely to rank public websites positively, while Republicans were more likely to rank them negatively. There were no significant differences, though, in assessments of government websites by sex, age, education, or race. This is noteworthy because it suggests that underlike some other political issues, demographic factors were not crucial in citizens' overall assessments of websites. There was no gender gap, race gap, education gap, or generation gap in how people evaluated the quality of government sites.

There is some tie between assessments of e-government and views about traditional government. If one does simple bivariate Pearson correlations between assessments of government website quality and three indicators of attitudes toward traditional government, there are significant associations with confidence in the federal government ($r = .15$; significant at .01; $N = 404$) and trust in government ($r = .11$; significant at .05; $N = 406$), but not the belief that government is an effective problem-solver ($r = .07$; not significant;

TABLE 7-1
Predictors of Citizen Evaluations of Government Website Quality, 2000

Sex	−.04 (.07)
Age	−.005 (.01)
Education	.02 (.02)
Party Identification	.04 (.02)*
Race	−.08 (.10)
Constant	2.08 (.24)
Adjusted R Square	.004
F	1.29
N	341

*$p < .05$

Source: Council for Excellence in Government National Survey, 2000

Note: The numbers are ordinary least squares regression coefficients with standard errors in parentheses. Tolerance statistics show no multicollinearity problem in the model.

$N = 403$). This means that those who are confident and trusting in traditional government also tend to be positive about e-government.

In part, citizens evaluate e-government based on how they feel about traditional government. There is a link between partisanship and feelings about e-government, exactly as one finds with bricks-and-mortar agencies. At the same time, individuals who mistrust government also tend to feel more negatively about e-government. If a person believes the public sector is undeserving of trust and confidence, those critical views carry over to the online world.

There are, however, other factors beyond e-government that are important to attitudes about traditional government, such as the quality of actual service delivery, views about government's proper role in society, and feelings about the ability of the public sector to bring about significant change. Most people still access government services in traditional sorts of ways, which means that effective, efficient, and responsive service delivery is more likely to be noticed among traditional than e-government. People who believe government has a constructive role to play in society are going to be more sympathetic to the public sector regardless of how they feel about e-government. To a certain extent, then, traditional government is evaluated on different terms than that of the digital variety.

CITIZEN USAGE

Beyond the issue of citizen sentiment about e-government is use of public sector websites. Citizen evaluations are affective measures that gauge overall feelings, but there needs to be behavioral measures of actual usage. How many people are going to government websites and who are the types of people relying on e-government?

According to the August 2000 national public opinion survey by the Council for Excellence in Government in Washington, D.C., only around half of Americans (54 percent) say they have visited a federal agency website. Even smaller numbers (45 and 36 percent, respectively) claim to have gone to a state or local government website.

These levels of e-government usage suggest that anywhere from half to two-thirds of the adult population remain outside the world of digital government. This poses a serious structural constraint to the ability of technology to transform citizen behavior and attitudes. E-government obviously cannot produce a major transformation in citizen usage as long as a majority of people are not accessing government websites.

There are some differences in usage levels by subgroup. An analysis of figures for federal, state, and local government websites demonstrates some important variations (table 7-2). For citizen usage of federal websites, age and education were the most significant predictors. Those who were young and well educated were the most likely to visit federal sites. The same was true for state government website visitors. With local government sites, however, the most significant predictors were education and race. Those who were well educated and/or part of a minority were the most likely to visit local websites.

The latter result is noteworthy because it shows a race gap in a direction that favors minority usage at the local level. Minorities are more likely to access government websites at the local, but not state or federal level because many minorities are concentrated in urban areas and therefore are more likely to be interested in what is on local government websites. In contrast to the prediction of a digital divide between whites and minorities, the latter will access sites if they believe it has content relevant to their lives.[31]

There have been some modest improvements over the course of time, mostly consistent with incremental models of change. A November 2001 and a February 2003 survey undertaken by the Council for Excellence in Government demonstrated some increases in citizen usage.[32] For example, citizen usage of state government websites rose from 45 percent in 2000 to 54 percent in 2001 and 54 percent in 2003. For local sites, usage increases from 36 percent in 2000 to 42 pecent in 2001 and 43 percent in 2003. And at the the federal level, usage rose from 54 percent in 2000 to 57 percent in 2001 and 59 percent in 2003.

There also were increases in the proportion of citizens reporting favorable views about e-government. In 2003, the percentage saying they believed e-government was having a positive impact on the way government operates was 58 percent, compared to 52 percent in 2001 and 35 percent in 2000.[33] Twenty-three percent indicated they were neutral, 9 percent said they were negative, and 10 percent were unsure. This suggests some improvement (but

TABLE 7-2

Predictors of Citizen Usage of Federal, State, and Local Government Websites, 2000

	Federal	State	Local
Sex	.05 (.04)	.05 (.04)	.07 (.04)
Age	.02 (.01)*	.02 (.01)*	.01 (.01)
Education	−.07 (.01)***	−.05 (.01)***	−.02 (.01)*
Party Identification	−.01 (.01)	−.01 (.01)	−.01 (.01)
Race	.07 (.06)	−.02 (.06)	.18 (.06)**
Constant	1.57 (.14)	1.68 (.15)	1.29 (.14)
Adjusted R Square	.07	.03	.03
F	8.58***	4.37***	3.87**
N	513	513	515

$^*p < .05;\ ^{**}p < .01;\ ^{***}p < .001$

Source: Council for Excellence in Government National Survey, 2000

Note: The numbers are ordinary least squares regression coefficients with standard errors in parentheses. Tolerance statistics show no multicollinearity problem in the model.

no systematic transformation) in optimism about how electronic governance is affecting public sector operations.

Overall, though, ratings of government websites in general were about the same in 2001 as in the year before. Sixty-nine percent rated the quality of public sector websites they had visited as excellent or good (down from 71 percent in 2000), 26 percent said they were only fair, 3 percent gave them poor marks, and 2 percent were unsure. The September 11, 2001, terrorist attacks on the United States did not harm citizen assessments of e-government websites across the country.

QUALITIES OF E-GOVERNMENT USERS

Aggregate patterns can conceal interesting subgroup differences below the surface. Therefore, in addition to looking at general results, it is important to break down e-government usage by various demographic categories. How does usage vary by gender, age, education, and income? And what do these patterns tell us about citizen evaluations of e-government?

In looking at subgroup differences, e-government users tended to be more male, younger, better educated, and earning higher incomes than the public as a whole. Consistent with an interpretation centering on a digital divide, there is a significant difference in e-government usage based on sex, age, education, and income. Older individuals, in general, make less use of the Internet so it is no surprise that they also are less likely to access e-government websites. Women and those with lower education attainment and lower incomes also are less likely to visit government webites due to lower level of interest in such sites and the existence of other kinds of demands that are placed on their time.

With the exception of age, these patterns are consistent with other forms of media usage and some types of political participation.[34] In general, people of higher education and income participate more, make greater use of information, and read or view media outlets more frequently. Men also have higher patterns of media consumption as a general pattern. The fact that e-government usage reflects other kinds of media consumption and political participation shows that many people do not see it as distinctive or revolutionary in nature but comparable to other information sources that already exist.

The only major exception to the typical profile is the relative youth of e-government users.[35] Young people are far more likely than older people to access the Internet in general and e-government in particular. This differs from the typical pattern with other media usage and political participation and obviously reflects the fact that young people often are more open to new technology, especially computer technology to which they are exposed at an early age in school. Even though their parents and grandparents are much more likely to vote and read newspapers, they are not as interested in and skilled at using Internet technology.

The interesting question is whether as youthful e-government users age and older nonusers pass away, there will be an exponential increase in e-government usage. It is conceivable that the distinctive age differences found in this area offer the possibility of skyrocketing usage over time, and of a population that is much more interested in digital service delivery and electronic interactions with government. The fact that young people are the most likely to access government websites and are generally favorable about e-government suggests considerable optimism about the future of this area. Over the course of the next few decades, the public should become even more receptive to e-government services as its users age and gain more experience at going online to access information and services.

CONCLUSION

To summarize, public opinion about e-government is quite relevant to the diffusion of innovation. Technologies are not going to spread and be in a position to transform society and government unless the public is receptive to the new invention, generally positive about its usage, willing to employ it themselves, and not limited by subgroup in terms of who can use the technology. If citizens are divided about the technology or there are large numbers of people who are not able to take full advantage of it, the ability of the public sector to integrate technology and use it as a force for long-term change will be limited.

Based on these criteria, e-government faces a generally positive future. Citizen interest in and use of e-government has been rising in recent years. There is no dramatic transformation in attitudes toward and usage of government websites, but each year the numbers inch upwards in an incremental fashion. Americans are favorable in their attitudes about e-government. Other than concerns about privacy, confidentiality, and security, the technology does not have major negatives. These attitudes mean that the public opinion climate for the spread of e-government is optimistic and likely to stay positive in the future. This is especially the case among young people, the age group most likely to use e-government and feel favorably toward it.

There remain interesting challenges, however, in terms of how people evaluate and use Internet technology in the public sector. Democrats are more likely than Republicans to evaluate e-government positively, and there are differences in usage related to age, gender, education, race, and income. These partisan differences represent a potential warning signal to e-government officials because they create the potential for conflicts over e-government to unfold along ideological lines. With members of the GOP generally being less sympathetic to e-government, there is the risk that IT issues might devolve into a more highly politicized environment comparable to what already exists surrounding issues such as health care and education.

If this happens, it will complicate the ability of government officials to move forward on technological change. Right now, administrators sell their program by emphasizing nonpartisan, technocratic reform. E-government will bring better service delivery in a more efficient and effective manner. This allows IT to appeal to a broad range of the political spectrum. If, however, the issue becomes overly politicized, it would call for very different strategies of coalition building beyond mere technocratic reform. Officials would have to present more partisan and ideological arguments in favor of e-government. These could be based on appeals to serve underprivileged populations (from liberals) or the need to focus on business applications (from conservatives). It would be a very different political environment for the construction of government technology policy.

The digital divide that exists in regard to e-government represents a major policy challenge for the public sector. This divide means that not all people are able to access and take advantage of the benefits of electronic technology.[36] Poor and uneducated citizens lack the resources to buy computers and gain access to the Internet. And those who have not been trained in using IT are in a weaker position to see its virtues. While there are some parallels between how citizens evaluate e-government and traditional government, there is evidence that voters see electronic service delivery differently and do not necessarily assess the two forms using the same criteria.

The big obstacle for policymakers is how to increase citizen usage of e-government and close the digital divide that exists across demographic categories. Since many people are unable to access the Internet due to their inability to purchase and own a computer or pay for Internet access, the quickest public policy action is to place general-access computers in schools, libraries, and government agencies that are available to the general public. That way, even if one lacks the personal means to access the Internet, one can take advantage of Internet technology to access government through other avenues. Some government agencies already have created public access rooms, in which citizens can visit government websites. Until personal incomes rise to such an extent that all can utilize the technology, placing computers in public areas would appear to be the most effective strategy for closing the digital divide and ensuring that everyone has access to electronic services. In the long run, improved accessibility will raise usage levels and provide a further basis for social and political transformation.

CHAPTER 8

Trust and Confidence in E-Government

E-GOVERNMENT IS BASED on the promise of better service delivery at lower cost to the taxpayer.[1] Through economies of scale that become possible through use of new technologies, digital delivery systems save money and in the long run produce substantial savings in public sector operations. Citizens can access information and services from their homes or offices and do so in a way that saves the government money. Not only will this technological revolution increase the personal convenience to members of the public and private sectors, it will also provide a means for more efficient and effective government operation.

The long-term hope is for a revolution not just in terms of service delivery but a fundamental shift in how citizens feel about government. If the public sector becomes more efficient, responsive, and effective, it may be possible for citizens to re-engage with government, become more confident about its performance, and be more likely to trust the public sector. Should these attitudes emerge, this would be a startling transformation in the way people react to government in general. Rather than doubting that the government is on their side, people would feel that the public sector was serving collective interests and that technology was helping the public sector become more efficient in the process.

If this transformation in public opinion takes place, it would represent a dramatic break with current public sentiment. As pointed out earlier, Americans have been extremely cynical about government for several decades. Public servants are seen as inefficient, ineffective, and unresponsive. Trust and confidence in government have dropped precipitously.[2] Public officials are seen as being little more honest and ethical than, for example, real estate agents, building contractors, and car salespeople.[3] Many citizens do not look to government for help but rather see it as a barrier to effective and efficient private action.

E-government proponents have suggested, however, that with the help of new technology, there is hope for turning around this cycle of cynicism.[4] Rather than assuming the government is bloated, wasteful, and inefficient, technology can improve public sector functioning, increase responsiveness, and deliver services in ways that are much more convenient to citizens. The whole idea behind electronic governance is that government can remedy long-standing societal perceptions about the public sector, and therefore alter fundamental citizen attitudes in the process.[5]

In this chapter, I look at this link between e-government and citizen trust and confidence in government. Is there a relationship between e-government usage and trust in government? Can increasing usage alter attitudes toward the public sector? Will the increasing popularity of digital service delivery change underlying citizen impressions about how public agencies function?

To examine this relationship, I analyze the 2000 national public opinion survey sponsored by the Council for Excellence in Government. This survey was undertaken to study the possibility that e-government usage might improve citizen attitudes about the public sector. Through questions about trust in government, assessments of government performance, and confidence in government, I demonstrate that there is reason for optimism about a revolution in citizen thinking about governmental agencies. Under certain circumstances, there is an association between visiting government websites and feeling more confident about government in general. When citizens were asked a series of "before and after" e-government questions, they shifted in the direction of believing that government was effective at solving problems.

This suggests that as people become more familiar with and comfortable using digital technology, their beliefs can be transformed in a more positive direction. As long as e-government avoids the partisan controversies and scandals of traditional government, there is hope of a public opinion transformation linked to electronic governance. As more money has flowed into e-government, however, there have been resulting partisan controversies and contracting scandals that undermine public confidence in electronic government. This is moving e-government from the realm of a neutral, technocratic policy reform to a partisan political issue.

TRENDS IN GOVERNMENT TRUST

Americans used to exhibit considerable trust that the government in Washington would do what is right.[6] In the 1930s, 1940s, and 1950s, the national government successfully won a world war and overcame the suffering of the Great Depression. People saw the public sector as dealing with important problems that were relevant to people and being honest in its approach to societal problem solving. Politicians were viewed as public servants who worked on behalf of the public interest, not special interest groups.[7]

The contrast with contemporary sentiments could not be more stark. In recent years, two-thirds of Americans say they mistrust the government in Washington.[8] They do not believe that public officials serve collective interests or are very honest in their dealings with outside interests. Citizens are more likely to believe public officials are serving the special interests instead of ordinary people. And as an occupational grouping, politicians are not seen as very honest or ethical. Rather, voters place them near the bottom of occupational inventories for personal integrity.[9]

Many problems have contributed to this rising tide of public mistrust. In the 1960s, citizens started to move in this direction during the Vietnam War. At that time, government officials told Americans the war was going well. United States troops were killing record numbers of Viet Cong troops. It appeared only a matter of time before the enemy was drained of energy and resources and America would win the war.

A number of news organizations sent reporters to the frontlines, however. These journalists saw a strikingly different picture. Rather than a war that was going well, the reporters witnessed one in which problems were readily apparent.[10] North Vietnamese troops were gaining ground and seemed to have substantial support among the general public. Instead of breaking their will, American bombing raids and military excursions appeared to make little headway.

Even worse, journalists discovered an unseemly side to American troop conduct.[11] Reports began to seep out showing that United States military personnel were committing atrocities. Innocent men, women, and children were being killed. As American troop losses reached into the tens of thousands, and massacres were documented, public support for the war started to drop.

As their patience for the war effort decreased, citizens also began to doubt the messenger. Trust in military and political officials dropped and people wondered why the government was lying to them. In the end, President Lyndon Johnson chose not to seek re-election. A new president, Richard Nixon, came into office. Over a period of several years, he pulled out American troops. By 1974, the North Vietnamese Army had captured control of South Vietnam, forced remaining U.S. personnel to leave, and unified their own country. It was a national trauma for the United States that undermined confidence in governmental performance.

Within a few years, President Nixon became engulfed in his own personal scandal. Seeking intelligence on the Democratic party, burglars funded by the Committee to Re-Elect the President broke into the Watergate office complex housing the Democratic National Committee. Caught red-handed in the office by security officers, Nixon's staff initially denied any White House involvement.

But between 1972 and 1974, a skeptical press, Congress, and court system combined to uncover information demonstrating that Nixon knew about the burglars, had funded their silence, and engaged in a massive cover-up to reduce the president's legal and political exposure.[12] Again, a number of government officials were found to have lied and President Nixon resigned from office rather than face possible impeachment by the U.S. House of Representatives. On the heels of the Vietnam debacle, the Watergate scandal further lowered public confidence in and trust toward the government in Washington.

For nearly a decade following these traumas, economic stagflation ravaged the United States. The economy performed poorly and the so-called "misery index" (the combination of inflation and unemployment) reached all-time highs. Public confidence in government plummeted during the 1970s.[13] Rising oil prices crippled the economy and virtually no one was happy with economic performance. Interest rates reached 20 percent, and both unemployment and inflation were in the double digits.

Seeking to pin the blame on the incumbent chief executive, Jimmy Carter, new President Ronald Reagan came into the White House and made a number of changes. He cut taxes, increased military expenditures, and adopted a tougher foreign policy. In the short run, it looked as if he would be able to turn around the slide in public confidence and make people feel more positive about the effectiveness of the public sector.

His closing years in office were tarnished by the Iran-Contra scandal, however. Rather than bucking the trend toward mistrust, this scandal reinforced an image of politicians who lied and attempted to cover up their misdeeds. Congressional hearings painted an unflattering image of foreign policy, one in which some officials had traded arms for hostages in Iran and Nicaragua. The entire episode had the predictable effect of undermining confidence in government.[14]

In the 1990s, Bill Clinton came into office hoping to move the Democratic party back into the political center and restore public trust in government officials. He set an ambitious course of remaking public policy. He proposed a health care plan that would provide health insurance for all Americans, make it easier to move from job to job without losing health coverage, and deal with the problem of rising health care costs. His plan went nowhere, and Republicans captured control of both the House and Senate in 1994.

In 1998, webcaster Matt Drudge broke a story claiming Clinton was having an affair with a White House intern named Monica Lewinsky. Although the president initially denied these charges, eventually he was forced to admit to a relationship with Lewinsky and, later, that he had not been completely forthcoming in his grand jury testimony about the relationship.[15] Clinton was impeached by the House, but the Senate chose not to remove him from office.

This litany of events over a period of three decades did little to make citizens have confidence in the ability of government agencies to solve problems and improve the lot of ordinary persons.[16] Instead, common folks associated the government with dishonesty, unethical personal behavior, and poor performance. Officials could not be trusted and were seen as doing more to help "big interests" than the "little guy." The cumulative impact of these examples of leadership deceit and poor performance reinforced negative perceptions about the public sector.

Following the tragic September 11, 2001, terrorist attacks on New York City and Washington, D.C., ordinary people associated the national government with more successful performance.[17] These attacks boosted trust in government back up to levels not seen in several decades. Voters rallied behind President George W. Bush and thought he was taking effective action against terrorists abroad. Not only did Bush's personal support rise, citizens generalized this positive feeling toward the government as a whole. People began to trust government officials to do what is right and to think they were fighting on behalf of collective interests.

But by the following year, these gains proved to be quite transient. Voters were about as mistrusting of government as they had been before the terrorist incidents. With a weak economy and partisan divisions over the Iraq war, many did not trust the government to handle major challenges and felt that reporters and politicians were not very honest or ethical.[18] Rather than seeing the public sector as effective in dealing with problems, citizens reverted to their previous cynicism. In the end, the September 11 public opinion rally in trust turned out to be a blip, not a new trend among the general public.

E-GOVERNMENT AND TRUST IN GOVERNMENT

Given public dismay with traditional government, it is an important question whether e-government usage is associated with higher public confidence in government. A 2000 national survey conducted by the Council for Excellence in Government was set up explicitly to test the relationship between e-government usage and public trust and confidence in government. One of its goals was to see whether the generally positive sentiments that people hold toward the Internet carries over to fundamental perceptions about government and the trustworthiness of public officials.[19]

If e-government is a tool for the transformation of service delivery, it is conceivable that there will be public opinion effects. Many of the activities that dragged down public confidence in government and made citizens see the public sector as inefficient, unresponsive, and ineffective are problems officials hope to remedy through electronic governance. Planners want to use technology to make government more responsive and to improve the functioning of government agencies.[20] If successful in reaching these objectives, there very well could be a tie between using e-government and overall attitudes toward the public sector.

To look at trust and confidence in government, we investigated the relationship between e-government usage and three key measures of attitudes toward and involvement with the public sector: trust in government, confidence in government, and belief that government is effective at solving problems and helping people.[21] These are basic attitudes and actions that are fun-

TABLE 8-1
Impact of E-Government Usage on Citizen Attitudes, 2000

	Trust in Government	Confidence in Government	Belief Government is Effective
Fed Govt Web Usage	.015 (.058)	.097 (.088)	−.092 (.078)
Party ID	.047 (.013)***	.087 (.020)***	.062 (.018)***
Education	−.0166 (.016)	.011 (.025)	−.002 (.022)
Age	.006 (.01)	.007 (.015)	.023 (.013)
Race	−.0099 (.08)	−.122 (.122)	−.161 (107)
Income	−0001 (.016)	−.045 (.024)	.014 (.021)
Sex	−.030 (.056)	−.033 (.085)	−.118 (.076)
Constant	2.63 (.22)***	2.81 (.33)***	2.70 (.29)***
Adjusted R Square	.024	.041	.040
F	2.43*	3.50*	3.40**
N	416	413	410

***$p < .001$; **$p < .01$
Source: Council for Excellence in Government National Survey, 2000
Note: The numbers are ordinary least squares regression coefficients with standard errors in parentheses. Tolerance statistics show no multicollinearity problem in the model.

damental to how people think about the public sector. How much trust and confidence do they have in government to solve problems? The more trusting and confident ordinary citizens are toward the public sector, the more favorable they feel toward government. Democracy depends on reasonable levels of citizen trust for its successful operation.[22]

Table 8-1 presents the results of an ordinary least squares regression analysis of federal e-government usage on these measures, controlling for party identification, age, education, race, sex, and income. We control for these political and demographic factors because in other work, they have been identified as relevant for attitudes about government and citizen activity levels. Party attachments, for example, structure people's general orientations toward government, while factors such as age, education, race, sex, and income often are associated with participation and trust in government.[23]

In looking at the results, there is no significant relationship between visiting federal government websites and views about trust, confidence, or government effectiveness.[24] E-government users are no more likely than nonusers to be trusting or confident about government or to believe the government is effective in solving problems. This suggests that at least so far, there is no revolution in citizen attitudes based on e-government usage. In contrast to rosy predictions from proponents, e-government is not associated with improving levels of trust and beliefs about the effectiveness of public sector problem solving.

Rather, the most significant predictor of these attitudes was party identification. Strong Democrats are the political grouping most likely to trust government, have confidence in government, and believe that government is effective at solving problems. In general, Democrats hold more favorable views about the public sector than Republicans, and partisan differences therefore are associated with a variety of attitudes about government functioning and performance.

A Priming Test

So far, this research has emphasized cross-sectional examination of the impact of e-government on citizen attitudes and behavior. Using data from a national survey, I have examined the degree to which patterns of e-government usage are associated with underlying feelings about government. The results do not suggest much of a tie between e-government reliance and trust in public sector performance.

A novel feature of this national survey, however, was that it included a component designed to compare citizen views on two sets of questions about government effectiveness and beliefs that e-government should have a high priority. At the beginning of the survey, questions were asked of each respondent regarding how high the priority should be for government to invest tax dollars in making information and services available over the Internet, and how effective government was. Then, after a series of e-government questions examining what services and information they would like to see online, how much usage has been made of government websites, and what they like and fear about e-government, respondents were asked an identical set of questions about government effectiveness and priorities.

This before-and-after design allows researchers to investigate questionnaire "priming," that is, change in each respondent's views after they have been exposed to a variety of e-government questions. The technique is a way to simulate the degree to which the introduction of detailed questions about e-government affects the potential to transform citizen beliefs. If citizens shift toward thinking government is more effective at problem solving after hearing about the possible benefits of e-government, it suggests at least the potential for e-government usage to boost citizen confidence in government.

For government effectiveness, respondents were asked the before-and-after item, "How effective do you think government is today at solving problems and helping people? (Very effective, fairly effective, fairly ineffective, or very ineffective)." For the e-government priority item, the survey used the before-and-after question, "In your view, how high a priority should it be for government to invest tax dollars in making information and services available over the Internet? (A very high priority, a high priority, a medium priority, a low priority, or a very low priority"). Subtracting the before measure from the after measure for each item created two scales dealing with individual-level attitude change. The *change in the e-government priority scale* ranged from –4 to +4 and measured movement toward (–4) or away from (+4) the view that e-government should be a high priority. The *change in government effectiveness scale* ranged from –3 to +3 and measured movement toward (–3) or away from (+3) the belief the government was effective. In this respect, therefore, these numerical scales represent an intraquestionnaire technique for measuring short-term attitude change following exposure to various kinds of e-government information.

As shown in table 8-2, an analysis demonstrates that there were significant changes in people thinking government was effective based on e-government usage and exposure to e-government questions. Those who visited federal government websites were more likely during the course of the questionnaire to move in the direction of feeling government was effective.[25] Citizens who were exposed to priming through e-government questions became more likely to express the opinion that government was effective at solving problems. Independently of whether they were Republicans or Democrats, or rich or poor (or in other demographic categories), there was an independent e-government effect on underlying confidence in government.

At the same time, federal government website visitors were not more likely to believe that e-government was a high priority for public sector tax investments. This finding is noteworthy because it is measuring budget receptivity as opposed to general orientations toward government problem solving. Befitting a question dealing with tax funding of new technology, traditional factors such as party identification (with Republicans more opposed than Democrats to spending on IT) constrained the ability of e-government usage to affect views about spending priorities.

In short, this "priming" experiment uses before-and-after questions to suggest that with educational effort about the issue of e-government, citizen beliefs can be transformed in a direction positive for beliefs about government effectiveness, but not spending priorities. As citizens become more informed about and familiar with e-government, public sector website usage has the potential to reshape underlying views about government. Those who make use of government websites are more likely to develop positive views about the public sector being effective at solving problems. The same conclusion, however, does not hold for judgments about spending priorities. Those beliefs are more deeply rooted and dependent on long-term feelings about government, such as citizens' partisan leanings. It is much more difficult to change these firmly held dispositions because they are passed down over a lengthy period of time.

TABLE 8-2
Impact of E-Government Usage on Changes in Citizen Attitudes, 2000

	Change in Seeing E-Government as Priority	Change in Seeing Government is Effective
Fed Govt Web Usage	−.199 (.104)*	.153 (.078)*
Party ID	.053 (.024)*	−.028 (.018)
Education	.016 (.030)	.0005 (.022)
Age	−.033 (.018)	−.014 (.013)
Race	−.088 (.146)	.062 (.107)
Income	.041 (.029)	−.012 (.021)
Sex	.011 (.101)	.029 (.076)
Constant	−.13 (.40)	−.21 (.293)
Adjusted R Square	.028	.007
F	2.70**	1.39
N	418	404

$^*p < .05$

Source: Council for Excellence in Government National Survey, 2000

Note: The numbers are ordinary least squares regression coefficients with standard errors in parentheses. Tolerance statistics show no multicollinearity problem in the model.

THE TRUSTWORTHINESS OF E-GOVERNMENT

While there is some encouragement in the results that e-government usage is associated with improvements in beliefs that government is effective, partisan controversies over e-government's development are undermining the sense that this area is nonpolitical and neutral. In the formative period of e-government, it was a "technocratic" reform that transcended party or ideological cleavages. Proponents sold new Internet technology as a nonpartisan way to promote greater efficiency and effectiveness in the public sector.[26] This nonpolitical vision helped generate public and political support, and thereby extend technology throughout government agencies.[27]

It is not clear, however, whether digital government can maintain its image as a nonpolitical, technocratic reform. When politicians start arguing about procurement and overspending by outside consultants, it becomes more difficult for the public to see it as a nonpartisan issue. Rather than being a device for revitalizing citizen trust in government, e-government will be seen as just another inefficiency in the public sector. To the extent that voters associate it with cost overruns, contracting scandals, and partisan controversies, e-government will lose its luster as a technocratic solution for improving government performance. The controversies described in chapter 2 surrounding e-government in Arkansas and Wisconsin suggest that the ability of electronic governance to restore public confidence in government may be limited.

Increasingly, financial and political scandals involving e-government are casting doubt on its long-term future.[28] As more and more public money flows into electronic government, public officials are relying on outsourcing and privatization. Most large-scale contracts for the development of website portals and electronic service delivery are being contracted to commercial firms, and some of these contracts are coming under close scrutiny.

In Detroit, for example, a contractor called Advanced Systems Resources got into political hot water when it billed the Detroit City Council $1.4 million for the reconstruction of its website. Council members alleged that costs rose out of control and that work was undertaken without proper authorization by city authorities. In the company's defense, company president Wendy Jackson said, "content management was something that was their responsibility, but we ended up taking it on, and that was the bulk of the project." Council member Sheila Cockrel complained that "we need to see a contract, there needs to be an audit, and it needs to provide a verification of what the deliverable [service] is for every bill that we are being asked to pay."[29]

In addition, there have been e-government scandals in California. Arun Baheti, the chief technology adviser to former Governor Gray Davis, resigned after he took a $25,000 campaign contribution for the governor's re-election from Ravi Mehta, an Oracle lobbyist, while the two were negotiating a $95 million software contract. News articles about the scandal raised the question of whether state officials did "favors for contributors" in order to help them gain government contracts. Previously, Baheti had attracted negative coverage when he ordered "100 state agencies to put Davis' photograph on their Web sites."[30]

A few weeks later, another California e-government official resigned amidst allegations of conflicts of interest over technology contracts. Companies employing relatives of top IT officials received state contracts of over $1 million. These charges led the state assembly to pass legislation requiring competitive bidding for IT contracts.[31]

Similar controversies befell a Utah chief information official named Phil Windley. He resigned after allegations of "favoritism" in the awarding of state technology contracts. According to news reports and a legislative investigation, contracts were awarded "under questionable circumstances." Legislators subsequently demanded an audit and formal legal investigation to determine if any laws were broken in the awarding of large state contracts.[32]

CONCLUSION

Not surprisingly, in light of such controversies, media coverage is starting to focus on performance issues and cost overruns associated with IT. For example, a 2003 *Business Week* article pointed out that many IT contractors in the public sector have failed to meet deadlines and have not devised systems that function effectively.[33] Among the specific cases cited were a Tennessee redesign of a Medicaid processing system, an $8 billion contract to overhall IRS computers, and a Georgia project to upgrade the state's health-processing technology. This undermines public support in new technology and willingness to pay for needed innovation.

For e-government to stay clean and retain its aura of a nonpartisan, technocratic reform, there must be no hint of corruption or scandal in the awarding of government contracts. Citizens must be reassured that e-government is different from traditional government. Contracting controversies damage electronic governance because they suggest that politicians are using e-government to reward friends and enhance their own political support.

For e-government to be successful at rejuvenating public trust, there must be an increase in citizen usage of e-government, citizen satisfaction with the service delivery that comes out of this interaction, and a reduction in the cost of government services. These are the types of things that proved problematic in the expansion of government during the twentieth century and helped to produce citizen skepticism. If government truly can deliver services at lower cost, that would be a powerful encouragement of trust and confidence in the public sector.

In the end, we should expect the public to be bottom-line oriented in its evaluation of digital government.[34] Evidence that confirms better service delivery at lower cost will go a long way toward restoring popular confidence in government. It would show that proponents were right to set lofty goals for themselves and that they then delivered concrete results based on these expectations. Anything that falls short of these expectations, though, will prove disheartening to average voters. If e-government does not deliver what its proponents claim, then it will appear as just another technology that promised much but was unable to deliver on all of its stated ambitions.

CHAPTER 9

Global E-Government

THE EXPERIENCE OF THE United States with respect to the way in which new technologies are incorporated into the public sector is revealing. Through a content analysis of government websites, examination of bureaucrat and public opinion surveys, case studies, budget expenditures, and aggregate statistical analysis, we have shown that e-government in the United States has developed in a largely incremental fashion. Change has focused on service delivery more than interactive democracy, and progress has emerged slowly over time.

Before generalizing this conclusion, though, it is important to look outside the United States. Thinking globally about technological change is useful because it broadens the scope to countries that have very different political, organizational, and financial dynamics. One of the limits of relying on any single country is the difficulty of generalizing beyond that area, especially to settings that feature different kinds of financial resources, political settings, or institutional structures.[1] Since these factors have been shown to be important for technology diffusion, an international approach gives researchers a chance to see how context affects the ability of countries to innovate in the technology area. In particular, it is important to look globally to determine whether e-government falls within models of incremental, secular, or transformational change. This will help us determine what is unique to the U.S. experience and what can be generalized more broadly.[2]

In this chapter, we look at global e-government in 2001, 2002, and 2003 to investigate the extent to which digital technology is producing a system-wide transformation or falling more within models of small-scale change predicted by secular or incremental perspectives. Using a detailed content analysis of 2,288 government websites in 196 different nations in 2001, a study of 1,197 websites in 198 nations in 2002, and investigation of 2,166 websites in 198 nations in 2003, we measure the information and services that are online and discuss how e-government varies by region of the world.[3] By focusing on all the countries around the world, this study adds a degree of generalizability that has been lacking in many other projects. We also analyze international public opinion surveys and an email responsiveness test to see how foreign governments compare to their U.S. counterparts. Finally, we undertake an aggregate, cross-national analysis to determine why some countries have made greater progress than others (modeled after the similar analysis undertaken of the fifty American states in chapter 4).

In general, we found that global e-government is not producing a major transformation of the public sector. While some countries have embraced digital government, a number of other countries have not placed much information or services online, and are not taking advantage of the interactive features of the Internet. Most nations have progressed no further than stage one (billboards) or stage two (partial service delivery) e-government. Furthermore, non-American websites are not very responsive to requests for information, even after controlling for language differences. The aggregate analysis of global e-government reveals that countries with limited wealth and low Internet usage do not have well-developed websites.

In addition, there is a weak link between democratization and e-government performance. Authoritarian governments are as likely to rank highly on e-government as democracies having free elections. In the long run, the openness and transparency of e-government may encourage the public sector to move towards democracy, but there is nothing immutable about this connection. Nondemocracies find it easy to issue top-down edicts that overcome bureaucratic intransigence, which helps them integrate technology into the public sector.

CROSS-NATIONAL COMPARISONS

It is difficult to compare countries around the world because of their sheer heterogeneity in terms of economic development, regime type, cultural patterns, telecommunications infrastructure, and Internet usage. Some countries are extremely rich, while many others lack the means to supply adequate education, housing, and health care. In terms of political structure, nations vary from presidential democracies and parliamentary systems to military dictatorships, monarchies, and authoritarian regimes. Openness to information technology fluctuates from areas where broadband is widely available to places where television and other forms of media transmission have been banned. Telecommunications structures vary from state monopolies to competitive market systems.

Considerable research has been undertaken on why some countries feature high Internet connectivity and others do not. In Lebanon and Russia, for example, less than ten percent of the population has any kind of access to the Internet, either through telephone modems or broadband connections. These figures contrast with industrialized nations, which have Internet access levels of 60–70 percent.[4]

In looking at the situation cross-nationally, academic scholars have reached very different conclusions concerning these disparities. One of the earliest studies was undertaken by Eszter Hargittai, who compared 18 countries that were members of the Organization for Economic Cooperation and Development (OECD).[5] She focused on Internet penetration, and found that its best predictors were economic wealth (measured by Gross Domestic Product [GDP] per capita) and telecommunications policy. Countries that were wealthy and had a competitive market structure featured higher levels of connectivity.

In contrast, Sampsa Kiiski and Matti Pohjola looked at Internet hosts per capita for 1995 to 2000 for 23 OECD countries and telephone access charges for 141 countries (including those outside the OECD).[6] As did Hargittai, they found that GDP per capita affected growth in computer hosting, but that the competitiveness of telecommunications markets had no independent effect on Internet penetration. Economic wealth and digital access costs mattered more than market structure or regulatory approach. Investment in education did not affect the results for OECD countries, but was important for the larger sample of OECD and non-OECD nations.

Pippa Norris, meanwhile, undertook a cross-national comparison of 179 countries.[7] She examined the relationship between a variety of social, economic, and political factors and the number of people online in each nation. Basically, she found that economic wealth (measured by GDP per capita) and the level of research and development spending were the best predictors of Internet usage. Neither education nor the level of democratization were significantly linked to citizen usage.

Consistent with the results for Internet usage in general, there is tremendous variation in e-government participation around the world. A 2001 public opinion survey of 29,077 people in 27 countries undertaken by Taylor Nelson Sofres, a consulting company, found dramatic variations in the percentage of each country's population that had accessed online government. The results ranged from a high of 53 percent in Norway, 47 percent in Denmark, 46 percent in Canada, 45 percent in Finland, and 34 percent in the United States to 3 percent in Turkey, Indonesia, and Russia, 5 percent in Poland and Lithuania, and 8 percent in Slovakia and Latvia. The average across the 27-nation survey was 26 percent.[8]

There were age and gender gaps in e-government usage around the world. Whereas 36 percent of those under the age of 25 reported using e-government in the past month, only 7 percent of those over 65 did so and 2 percent of those aged 55 to 64. Twenty-eight percent of men used the Internet, compared to 24 percent of women. These numbers demonstrate that the division in access to e-government that exists in the United States is present elsewhere, as well. Many people do not have quick, easy, or convenient access to the Internet.

This same survey revealed interesting differences between the U.S. and non-U.S. experiences. Two-thirds of the people in the surveyed countries felt "unsafe" using the Internet to conduct government transactions online. These were individuals who worried about the overall security, privacy, and confidentiality of public sector websites. Germany was the nation having the highest percentage (85 percent) of people saying they felt unsafe, compared to 84 percent in Japan, 84 percent in France, 72 percent in the United States, and 72 percent in the Czech Republic. Countries that showed the lowest levels of unsafe feelings in regard to online government transactions were Estonia (27 percent), Canada (30 percent), Denmark (31 percent), Hong Kong (32 percent), and Finland (37 percent). Those reporting the greatest concern over Internet safety were women and people aged 25 to 44 years old.[9]

A year later, when Taylor Nelson Sofres completed a follow-up survey of 28,952 people in 31 countries, the percentage of the population in these countries that had used online government rose from 26 to 30 percent. There also was little change in what uses people made of online government. Twenty-four percent of adults reported they used e-government for information seeking, up from 20 percent in 2001. Eleven percent said they used it for downloading forms, up from 9 percent the previous year. Eight percent said they used e-government to provide information to the authorities, up from 7 percent in 2001. Seven percent indicated they provided bank account or credit card information to a government site, up from 6 percent the preceding year. Four percent said they used government websites to express a point of view.[10]

These results are consistent with a model of incremental change. Rather than seeing dramatic changes from year-to-year in e-government usage around the world, the report's authors concluded that "there has been very little change to the shape of the Government Online adoption curve." Both in terms of the percentage of people visiting government websites and the types of usage they made of these sites, there has been no dramatic transformation in citizen behavior.[11]

There were some changes over time in perceptions about global e-government from the citizen point of view. The percentage of people feeling it was okay to provide personal information to a government website rose from 14 percent in 2001 to 23 percent in 2002. While this demonstrates some improvement, the figures reveal that large numbers report feeling either unsafe or unsure about confidentiality when it comes to government websites.[12] Indeed, in some countries, large numbers of citizens report concerns about giving personal information online at government websites. For example, 90 percent of Japanese citizens say they are concerned about providing personal information, while 82 percent feel this way in Germany, 76 percent say this in France, 75 percent believe this in Taiwan, and 72 percent express this opinion in Italy.[13]

In the case of e-government, then, there is tremendous variation around the world in the extent to which governments have embraced the digital revolution.[14] Some are enthusiastic about e-government and have placed some services online (consistent with stage two e-government), while others remain mired in the billboard stage of e-government. In the latter countries, few materials are online other than reports and databases, and there is little attention to privacy or security.

There are obvious reasons for this disparity. Few nations have the financial resources, organizational capacity, and political will to make technology change a top priority. Resources vary enormously as do organizational capacity and political leadership. Countries that lack wealth have a difficult time justifying investment in new technology.[15] Nations where citizens, organizations, and politicians doubt the potential of technology face barriers to the implementation of electronic governance.

Research by Todd LaPorte and Chris Demchak, for example, found an association between a country's national income and the openness of its government websites. In particular, they found that e-government progress is related to the magnitude of capital flows across countries.[16] Nations that have access to capital are in a much stronger position to innovate than those that do not.

In her international study of online government, Norris found that "more affluent industrialized economies characteristically have the richest access to multiple forms of communication and information technologies . . . and this environment is most conducive to the spread of online parties as well."[17] This research demonstrates that wealth and political development matter to the incorporation of technology into political activities. Countries that rank highly on economic and political indicators show much stronger commitment to e-government and are in a much stronger position to execute its implementation in the public sphere.

What most of these studies have not addressed, however, is what is online at national government websites around the world and how quickly e-government has changed from year to year. Building on the methodology of the U.S. study, I supervised a team of researchers who examined the national government websites in 2001, 2002, and 2003 for all the nations around the world. We looked for material on information, services, and databases, features that would facilitate e-government access by special populations such as the disabled and non-native language speakers, interactive features that would facilitate outreach to the public, and visible statements that would reassure citizens worried about privacy and security over the Internet. We looked at a wide variety of political and economic systems, from monarchies, federated systems, and presidential democracies to parliamentary systems, dictatorships, and communist countries.

The data for the 2001 analysis consisted of 2,288 national government websites for the 196 nations around the world.[18] Among the sites analyzed were those of executive offices (such as a president, prime minister, ruler, party leader, or royalty), legislative offices (such as Congress, Parliament, or a People's Assembly), judicial offices (such as major national courts), cabinet offices, and major agencies serving crucial functions of government, such as health, human services, taxation, education, interior, economic development, administration, natural resources, foreign affairs, foreign investment, transportation, military, tourism, and business regulation. Websites for subnational units, obscure boards and commissions, local government, regional units, and municipal offices were not included in this study.[19]

The regional breakdowns for the websites studied in 2001 were 25 percent from Western or Eastern European countries, followed by 18 percent from Africa, 14 percent from Asia, 9 percent from Central America, 8 percent from the Middle East, 7 percent from Russia and Central Asia (such as the areas of the former Soviet Union), 7 percent from South America, 7 percent from the Pacific Ocean countries (meaning those off the continent of Asia), and 5 percent from North America (which included Canada, the United States, and Mexico).

In 2002, our research team looked at 1,197 government websites in 198 different nations. Regional breakdowns for the websites studied that year included 24 percent from Western European countries, followed by 17 percent from Asia, 14 percent from Africa, 8 percent from Eastern Europe, 8 percent from the Middle East, 7 percent from Central America, 7 percent from Russia and Central Asia (such as the areas of the former Soviet Union), 6 percent from North America (which included Canada, the United States, and Mexico), 4 percent from South America, and 4 percent from the Pacific Ocean countries (meaning those off the continent of Asia).

In 2003, we studied 2,166 websites in 198 countries. The regional breakdowns included 21 percent from Western European countries, followed by 17 percent from Africa, 12 percent from Asia, 11 percent from Eastern Europe, 8 percent from the Middle East, 8 percent from South America, 7 percent from the Pacific Ocean countries (meaning those off the Asian continent), 6 percent from Central America, 6 percent from North America (which included Canada, the United States, and Mexico), and 5 percent from Russia and Central Asia (such as the areas of the former Soviet Union).

National government websites reflected the social, economic, political, and religious background of a particular area. Muslim countries often had links to religious unity pages or offered forums where visitors could discuss religious issues. In some former communist nations, ministries of privatization aimed at foreign investors appeared to be the most elaborate sites. Nations that relied heavily on tourism (such as those in the Caribbean or Pacific islands) often centered their e-government activities around tourism sites.

Regardless of the type of system or cultural background of a country, websites were evaluated for the presence of more than two dozen different features dealing with information availability, service delivery, and public access. Features assessed included type of site, name of nation, region of the world, office phone number, office address, online publications, online database, external links to nongovernmental sites, audio clips, video clips, non-native languages or foreign language translation, commercial advertising, user payments or fees, subject index, disability access, privacy policy, security features, presence of online services, number of different services, links to a government services portal, digital signatures, credit card payments, email address, search capability, comment form or chat-room, broadcast of events, automatic email updates, personal digital assistant (PDA) accessibility, and having an English version of the website.

Where national government websites were not in English, our research team employed foreign language readers who translated and evaluated national government websites where possible. In some cases, we have made use of foreign language translation software available online through http://babelfish.altavista.com. This allowed us to assess government websites in the native language of a particular country.

GLOBAL E-GOVERNMENT TRENDS

Several patterns stand out in the study of global e-government. One is the extent to which English has become the language of global e-government. Seventy-four percent of national government websites in 2003 had an English version, similar to the 78 percent in 2002 and the 72 percent in 2001. Reflecting the multilinguistic nature of global interactions, many nations offer more than one language on their websites. For example, around half (51 percent in 2003, 43 percent in 2002, and 45 percent in 2001) have two or more languages on their government sites. Other than English, common languages included Spanish, French, Russian, German, Italian, Portuguese, Arabic, and Chinese. By foreign language feature, we mean any accommodation to the non-native speakers in a particular country, such as text translation into a different language.

In terms of information availability, many countries have made progress from 2001 to 2003 at putting publications, forms, and databases online for citizen access. Government agencies have discovered that it is very efficient for the general public to be able to download common documents rather than having to visit or call the particular agency. Many countries, however, have not made similar progress in placing official government services online. There is wide variation across countries and by region of the world in the extent to which citizens access government services through the Internet. While some governments offer a number of services online, most do not.

Many countries do not have portals that link the services of various agencies and departments of that country. Portals offer many advantages for government offices. Having a single entry point into a national government helps citizens because these portals integrate e-government service offerings across different agencies. These central sites reduce the need to log on to different agency websites to order services or find information. Instead, citizens can engage in "one-stop" shopping, and find what they need at a single source. Service portals improve citizen access because they encourage more uniform designs for particular countries. Rather than have a "Tower of Babel" composed of different government agencies whose websites do not share a common navigational system, presentational style, or method of organization, these one-stop portals make it much easier for citizens to access online information and services.

As discussed later in this chapter, there remains a need for continuing advancement in the areas of privacy, security, and interactive features, such as search engines. Compared to various commercial websites, the public sector in many nations lags behind the private sector in making full use of the technological power of the Internet to improve the lives of citizens and enhance the performance of governmental units. Given public concerns about privacy and security on the Internet, governmental agencies need to do more to reassure the public that e-government is safe and secure for all users.

ONLINE INFORMATION

In looking at specific features of government websites, we wanted to see how much material was available that would help citizens contact government agencies and navigate websites. In general, contact information is quite prevalent, and there were improvements between 2001 and 2002 (this item was not analyzed in 2003). The vast majority of sites in 2002 provided their department's telephone number (77 percent) and mailing address (77 percent). These are materials that would help an ordinary citizen needing to contact a government agency reach that office.

In terms of the content of online material, many agencies have made extensive progress at placing information online for public access (table 9-1). In 2003, 89 percent of government websites around the world offered publications that a citizen could access, up from 77 percent in 2002, and 71 percent in 2001. Seventy-three percent offered databases, down from 83 percent in 2002 but up from 41 percent in 2001. Eighty-two percent had links in 2002 to external, nongovernmental sites where a citizen could turn for additional information, which is up from 42 percent in 2001.

As a sign of the early stage of global e-government, most public sector websites do not incorporate audio clips or video clips on their official sites. Despite the fact that these are becoming much more common features of e-commerce and private sector enterprise, only 8 percent of sites in 2003 (the same as in 2002, but up from 4 percent in 2001) provided audio clips, while 8 percent had video clips (down from 15 percent in 2002, but up from 4 per-

TABLE 9-1
Percentage of Global Government Websites Offering Publications and Databases, 2001–2003

	2001 (%)	2002 (%)	2003 (%)
Telephone Contact Information	70	77	—
Address Information	67	77	—
Links to Other Sites	42	82	—
Publications	71	77	89
Databases	41	83	73
Audio Clips	4	8	8
Video Clips	4	15	8

Source: Author's e-government content analysis database

cent in 2001). A common type of audio clip was a national anthem or a musical selection.

ONLINE SERVICES

Fully executable, online service delivery benefits both government and its constituents. Of the websites around the world, however, 16 percent offer services that are fully executable online, which is up from 12 percent in 2002 and 8 percent in 2001 (table 9-2). The slow expansion in the rate of service increase and the fact that most sites offer no services demonstrates that global e-government remains stuck in either stage one (billboards) or stage two (partial service delivery) e-government. Of the 2003 sites, 9 percent offer one service, 3 percent have two services, and 4 percent have three or more services. Eighty-four percent have no online services.

North America (including the United States, Canada, and Mexico) is the region of the world offering the highest percentage of online services (table 9-3). Forty-five percent (up from 41 percent in 2002 and 28 percent in 2001) had fully executable, online services. This was followed by Asia (26 percent), the Middle East (24 percent), the Pacific Ocean islands (17 percent), and Western Europe (17 percent). Only 1 percent in Russia/Central Asia, 5 percent in Africa, and 6 percent of sites in Eastern Europe offer online government services.

Of the 198 nations analyzed, there is wide variance in the number of online services provided by different governments. As shown in table 9-4, the country with the largest number of services is Singapore, with an average of 7.8 services across its government agencies. This is followed by the United States (4.8 services), Turkey (3.2 services), Hong Kong (3.1 services), and Tai-

TABLE 9-2
Percentage of Global Government Online Services, 2001–2003

	2001 (%)	2002 (%)	2003 (%)
None	92	88	84
One	5	7	9
Two	1	2	3
Three or more	2	3	4

Source: Author's e-government content analysis database

TABLE 9-3
Percentage of Global Government Sites Offering Online Services by Region of World, 2001–2003

	2001 (%)	2002 (%)	2003 (%)
North America	28	41	45
Pacific Ocean Islands	19	14	17
Asia	12	26	26
Middle East	10	15	24
Western Europe	9	10	17
Eastern Europe	—	2	6
Central America	4	4	9
South America	3	7	14
Russia/Central Asia	2	1	1
Africa	2	2	5

Source: Author's e-government content analysis database

TABLE 9-4
Average Number of Online Servives across Each Country's Websites, 2003

Singapore	7.8	Philippines	0.8
United States	4.8	Oman	0.7
Turkey	3.2	Switzerland	0.7
Hong Kong	3.1	Venezuela	0.7
Taiwan	2.4	Spain	0.7
Bahrain	1.8	Yemen	0.6
Saudi Arabia	1.2	Gambia	0.6
China	1.2	Great Britain	0.6
Guinea-Bissau	1.0	Canada	0.6

Source: Author's e-government content analysis database

wan (2.4 services). It is important to keep in mind that our definition of services included only those services that were fully executable online. If a citizen has to print out a form and mail or take it to a government agency to execute the service, we do not count that as an online service.

The most frequent services found included ordering publications online, buying stamps, filing complaints, applying for jobs, applying for passports, and renewing vehicle licenses. Several countries had novel online services. For example, the Dominican Republic's National Drug Control office had a "drug information" link in which anonymous citizens could report illegal drug dealing. Australia offered the possibility of applying for jobs online at some national agencies. Bangladesh's National Tourism Organization offered online booking of hotel rooms. Canada offers a number of services online such as change of postal address forms, package tracking, and ordering stamps. Egypt allows for personal and union registration online at the Ministry for Manpower and Emigration. Lithuania offers searches for stolen vehicles, invalid identity documents, and wanted persons through its Ministry of the Interior.

One of the features that has slowed the development of online services has been an inability to use credit cards and digital signatures on financial transactions. Of the government websites analyzed, however, only 2 percent in 2003 accepted credit cards and 0.10 percent allowed digital signatures for financial transactions (similar to the figure for 2002). Among the sites having a capacity for digital signatures are the Singapore governmental office of statistics and Denmark's portal site. Since some government services require a fee, not having a credit card payment system makes it difficult to place government services that are fully executable online and limits the transformational potential of the Internet.

The other limitation has been financial and political resources. Online services are expensive to develop. Since each country has different laws and regulations, there are few off-the-shelf software packages that are cheap and readily available. The need for unique programs obviously restricts the ability of poor nations to place services online for their citizens and business community. In addition, online services require a strong political will in order to get disparate agencies to work together. Top leaders must be committed to e-government development and must be willing to marshal the resources necessary for progress in this area. When a country has only a very small number of citizens with access to the Internet, it is difficult to justify the expenditure of scarce resources for online government.

PRIVACY AND SECURITY

The unregulated and accessible structure of the Internet has prompted many to question the privacy and security of government websites. National public opinion surveys place these areas near the top of the list of citizen concerns about e-government.[20] Having visible statements outlining what the site is doing to preserve user privacy and security are valuable assets for reassuring a fearful population and encouraging citizens to make use of e-government services and information.

Few global e-government sites, however, offer policy statements dealing with these topics. Only 12 percent in 2003 (similar to the 14 percent in 2002 and the 6 percent in 2001) had some form of privacy policy on their site, and 6 percent had a visible security policy (similar to the 9 percent in 2002 and 3 percent in 2001). Both of these are areas that government officials need to take much more seriously. Unless ordinary citizens feel safe and secure in their online information and service activities, e-government is not going to grow very rapidly. The slow rate of change in this area demonstrates that global e-government is moving in an incremental, not transformational, direction.

Despite the importance of security in the virtual world, there are wide variations across nations in the percentage of websites showing a security policy. The countries most likely to show a visible security policy are Singapore (90 percent of its sites), Canada (65 percent), and the United States (62 percent). This was followed by Australia (39 percent), New Zealand (30 percent), St. Lucia (25 percent), Great Britain (21 percent), Japan (15 percent), and Taiwan (12 percent). Most other nations do not have sites with a security statement.

Similar to the security area, there are widespread variations across the nations in providing privacy policies on their websites. The countries with the highest percentage of websites offering a visible privacy policy are Australia and Dominica (each with 100 percent of its sites). These nations were followed by Canada (97 percent), Singapore (93 percent), China (83 percent), United States (75 percent), St. Lucia (50 percent), New Zealand (47 percent), Great Britain (45 percent), and Taiwan (42 percent).

It is interesting that some of the countries that have the greatest number of privacy statements are nondemocratic regimes. Countries such as Singapore and China are not very open in their political systems, yet are paying attention to privacy on their government websites. As I discuss later in this chapter, this suggests a weak link between democratization and some aspects of e-government performance.

DISABILITY ACCESS

Disability access is vitally important to citizens who are hearing impaired, visually impaired, or suffer from some other type of disability. If a site is ill-equipped to provide access to individuals with disabilities, it fails in its attempt to reach out to as many people as possible. Using the automated online "Bobby" service at http://bobby.watchfire.com, we tested actual accessibility through the Priority Level One standards recommended by the World Wide Web Consortium (W3C). Sites were judged to be either in compliance or not in compliance based on the results of this test.

According to this test, 14 percent of government websites around the world were accessible to the disabled in 2003. This is far lower than comparable numbers for the United States national government (47 percent), U.S. state governments (33 percent), and U.S. local governments (20 percent). The low performance on this feature means that globally, special-needs populations are not receiving much help in terms of accessibility. Foreign governments do not appear to see disability access as a high priority for government investment.

ADS, USER FEES, AND PREMIUM FEES

Many nations are struggling with the issue of how to pay for electronic governance, yet use of advertisements to finance government websites is not very prevalent. Only 2 percent of sites in 2003 had commercial advertisements on their sites, meaning nongovernmental, corporate, and group sponsorships. This was down from 8 percent in 2002 and 4 percent in 2001. There also was little reliance on user fees (0.2 percent) or premium fees (0.2 percent) as of 2003, similar to the situation previous year. The countries with the greatest reliance on ads include Sri Lanka, Tuvalu, Bhutan, Antigua, and Guinea-Bissau (each with 100 percent of its government websites having ads). This is followed by St. Vincent (50 percent), Belize (50 percent), and Russia (37 percent).

Tourism was the government sector most likely to have advertisements. For example, these websites had banners or fly-by ads for hotels, travel agents, or special travel packages. Examples of ads included the Laos Ministry of Tourism (hotel booking services), Maldives Tourism (advertisement for advertisers), Paraguay Tourism (Portugal Investment portugalinbusiness.com pop-up), Russia's Tourist Office (Rambler's Top 100, spylog.com, and love boat singles cruises), Uruguay Tourism (vacation planning web sites with large banner ads for vende Uraguay), Belize Tourism (Radisson, KPMG, law offices), Antigua and Barbuda Portal/Tourism (TNT Vacations, Jolly Roger Pirate Ship banner ads), and Iran Tourism (travel agencies, private e-greetings site).

Other examples of public sector website advertisements included Mongolia Foreign Affairs (mongolmedia.com), Pakistan Railway (advertisements for advertisers), Russian Federation Departments of Agriculture, Information, Economy, Finance, Foreign Affairs, and Railway Transport), Saint Vincent and the Grenadines National Broadcasting (newmontrosehotel.com), Tuvalu Portal (allcasinoworld.com and jane's oceana page), Uruguay Portal/President (ITUraguay.com—"It's business. It's Uraguay."), Vietnam Finance (accounting firm banner ad), Vietnam News Agency (magazine and newspaper advertising and econet banner ads), Yugoslavia Investment/Export (real estate services, KPMG banner ads), Bhutan Portal (tourist agencies), Taiwan Transport/Communications (tourist agencies and television station), Germany Environment (banner ad), Guinea-Bissau Portal (Intership Limited shipping company and WorldNews ads), Kazakhstan Economy (geocities pop-up), Kenya Investment Promotion (search-related ads), Korea Portal (correctkorea.net, learn to speak Korean online, travel agency banner ads), and Kyrgyzstan Mineral Resources (jewellernet.ru).

For user fees, the nation with the greatest employment is Taiwan, with 8 percent of its sites having user fees. Other nations relying on user fees are Oman (5 percent of its sites), Singapore (3 percent), and Switzerland (3 percent). The only countries having premium fee areas are Afghanistan (100 percent), Barbados (25 percent), Taiwan (8 percent), and Canada (3 percent).

RESTRICTED AREAS

Some countries have started to develop restricted areas on their websites that require a username and password for accessibility. This can be for security reasons or because of an interest in personalizing service delivery. In 2003, 6 percent of government websites across the world had restricted areas, the same as the previous year.

Examples of countries with restricted areas on their sites include the Congo (100 percent of its sites), Madagascar (50 percent), Cook Islands (50 percent), Taiwan (42 percent), China (33 percent), Kuwait (33 percent), United States (30 percent), Oman (26 percent), Laos, Bahrain, Barbados, Belize, and Iran (each with 25 percent).

Examples of website restrictions included access to bulletin boards, forums, and newsgroups (Algeria, Chad, Taiwan, Malaysia, and Mexico), transportation and accommodation reservations in Taiwan, Ireland, and the Dominican Republic, the Resource Center on the Malaysia Trade and Industry page, and Intranets on the pages of Venezuela, Canada, Guatemala, and Indonesia.

PUBLIC OUTREACH

As pointed out in chapter 6, e-government offers the potential to bring citizens closer to their governments. Regardless of the type of political system that a country has, the public benefits from interactive features that facilitate communication between citizens and government. In our examination of national government websites, we looked for various features that would help citizens contact government officials and make use of information on websites.

TABLE 9-5
Percentage of Global Government Websites Offering Public Outreach, 2001–2003

	2001 (%)	2002 (%)	2003 (%)
Email	73	75	84
Search	38	54	—
Comments	8	33	31
Email Updates	6	10	12
Broadcast	2	2	—
Website Personalization	—	1	1
PDA Access	—	—	2

Source: Author's e-government content analysis database

Email is an interactive feature that allows ordinary citizens to pose questions of government officials or request information or services. As shown in table 9-5, we find that 84 percent of government websites in 2003 (up from 75 percent in 2002) offered email contact material so that a visitor could email a person in a particular department other than the Webmaster.

While email is the easiest method of contact, there are other methods that government websites employ to facilitate public feedback. These include areas to post comments, the use of message boards, and chat-rooms. Websites using these features allow citizens and department members alike to read and respond to others' comments regarding issues facing the department. This technology is less prevalent than email, with 31 percent of websites offering this feature in 2003, while 33 percent provided it in 2002.

Twelve percent (up from 10 percent in 2002) of government websites allow citizens to register to receive updates regarding specific issues. With this feature, web visitors can input their email addresses, street addresses, or telephone numbers to receive information about a particular subject as new information becomes available. The information can be in the form of a monthly e-newsletter highlighting a prime minister's views or in the form of alerts notifying citizens whenever a particular portion of the website is updated. One percent of sites allow websites to be personalized to the interests of the visitor, and 2 percent provide PDA access.

With the exception of email, the limited use of interactive features that facilitate citizen access and feedback shows that technological change has not advanced very far on the global scene. Most countries have not embraced a vision of e-government that sees it as a tool for citizen empowerment. Instead, officials view the Internet as a billboard for one-way communication with the public. They are not taking advantage of two-way features that provide citizens with a chance to voice their opinions or personalize websites to their particular interests.

EMAIL RESPONSIVENESS

It is useful to have email contact information on government websites, but this material is not helpful unless there is someone who actually answers the email. Responsiveness is one of the key claims by advocates for technological change. Internet proponents argue that this technology will make it easier for government officials to respond to citizen complaints and concerns.

In order to test how responsive various governments were to citizen inquiries, we sent email messages in 2002 to each of the 1,197 international government websites we assessed. Our message was a simple question in English: "I would like to know what hours your agency is open during the week. Thanks for your help." We tracked whether agencies responded, and if so, how many business days it took them to respond.

Only 19 percent of agencies responded to our question, 75 percent did not, and 6 percent had broken email links or addresses that prevented a response. Twelve percent responded within one day, 3 percent took two days, 2 percent responded in three days, and 2 percent replied in four or more days. These response rates are far lower than the U.S. case. In the United States, we generated responses from 68 percent of officials in 2003, 35 percent in 2002 (when there was a more challenging question to answer), 80 percent in 2001, and 91 percent in 2000. This demonstrates the extent to which international officials are less responsive to requests than is true in the United States.

Of course, language is one possible explanation for these results. Our question was sent in English, and it is possible that non-English-speaking countries chose not to reply due to language difficulties as opposed to an unwillingness to respond. To check this possibility, we cross-tabulated responsiveness by whether the country's website was in English or not. If the country had government websites in English, language difficulty would not explain the poor response rates.

Table 9-6 shows that although responses were lower from countries without English websites, the differences were not overwhelming. Twenty-two percent of English sites responded to the email request, compared to 10 percent of non-English websites. This suggests language was not the problem so much as an unwillingness to respond to citizen requests for information. Rather than becoming a mechanism to bring citizens closer to leaders, government officials are not using the Internet to take full advantage of new technology. Until this reality is changed, it will be difficult for new technology to improve government performance.

OVERALL E-GOVERNMENT PERFORMANCE

In order to see how the 198 nations ranked overall, we created a 0–100 point e-government index and applied it to each nation's websites based on the

TABLE 9-6
Percentage for Email Responsiveness by Whether Website was in English Language, 2002

	In English (%)	Not in English (%)
No Response	73	83
Responded in One Day	13	7
Responded in Two Days	4	2
Responded in Three Days	2	0
Responded in Four Days	1	0
Responded in Five Days	0	1
Responded in Six Days or More	2	0
Broken Email Link	5	7

Source: Author's e-government content analysis database

availability of publications, databases, and number of online services. Four points were awarded to each website for the presence of the following features: publications, databases, audio clips, video clips, foreign language access, not having ads, not having premium fees, not having restricted areas, not having user fees, disability access, having privacy policies, security policies, allowing digital signatures on transactions, an option to pay via credit cards, email contact information, areas to post comments, option for email updates, option for website personalization, and PDA accessibility. These features provide a maximum of 76 points for particular websites.

Each site then qualified for a bonus of 24 points based on the number of online services executable on that site (1 point for one service, 2 points for two services, 3 points for three services, and on up to twenty-four points for 24 or more services). The e-government index runs along a scale from 0 (having none of these features and no online services) to 100 (having all features plus at least 24 online services). Totals for each website within a country were averaged across all of that nation's websites to produce a zero–100 overall rating for that nation.

The top country in our ranking was Singapore at 46.3 percent.[21] This means that every website we analyzed for that nation had nearly half of the features important for information availability—citizen access, portal access, and service delivery. Other nations that scored well on e-government include United States (45.3 percent), Canada (42.4 percent), Australia (41.5 percent), Taiwan (41.3 percent), Turkey (38.3 percent), Great Britain (37.7 percent), Malaysia (36.7 percent), the Vatican (36.5 percent), and Austria (36.0 percent). Appendix II lists e-government scores for each of the 198 countries in 2003.

There were some differences in e-government by region of the world (table 9-7). In looking at the overall 2003 e-government scores by region, North

TABLE 9-7
Global E-Government Ratings by Region, 2001–2003

	2001 (%)	2002 (%)	2003 (%)
North America	51.0	60.4	40.2
Western Europe	34.1	47.6	33.1
Eastern Europe	—	43.5	32.0
Asia	34.0	48.7	34.3
Middle East	31.1	43.2	32.1
Russia/Central Asia	30.9	37.2	29.7
South America	30.7	42.0	29.5
Pacific Ocean Islands	30.6	39.5	32.1
Central America	27.7	41.4	28.6
Africa	23.5	36.8	27.6

Source: Author's e-government content analysis database

America scores the highest (40.2 percent), followed by Asia (34.3 percent), Western Europe (33.1 percent), Pacific Ocean islands (32.1 percent), Middle East (32.1 percent), Eastern Europe (32.0 percent), Russia and Central Asia (29.7 percent), South America (29.5 percent), Central America (28.6 percent), and Africa (27.6 percent).

In looking at regional differences by particular feature (table 9-8), North America, Asia, and the Middle East ranked most highly on services, while North America, Russia, and South America scored highest on access to databases. The areas providing the greatest degree of accessibility through PDAs was the Middle East.

EXPLAINING GLOBAL E-GOVERNMENT

To this point, we have described the extent of global e-government performance, but not explained variation across countries. In this section, we present an aggregate analysis of e-government. Using a conceptual model based on organizational, fiscal, and political factors, we seek to determine what distinguishes stronger e-government countries from those that are weaker.

For my measures of e-government performance, we rely on five indicators from the 2003 content analysis previously reported: number of online services, percent of each country's agencies that provide online services, the quality of a country's website privacy policies, its overall e-government rank, and its overall e-government numerical score. The number of online services is the average number of fully, executable electronic services across each of its specific agencies. The percentage of agencies offering on-

TABLE 9-8
Global E-Government Features by Region, 2003

	North America	Central America	South America	Western Europe	Eastern Europe	Russia	Middle East	Africa	Asia	Pacific Ocean Islands
Publication	96	87	94	96	92	98	80	79	90	82
Database	87	77	83	78	74	85	56	60	78	72
Audio Clip	18	10	7	6	6	4	10	6	10	7
Video Clip	17	4	8	9	7	4	8	2	18	3
Foreign Language	26	9	8	64	100	60	78	33	67	25
Ads	1	4	2	0	1	10	2	1	5	1
Premium Fee	1	1	0	0	0	0	1	0	1	0
Restricted Access	17	5	5	3	3	2	16	4	13	5
User Fee	0	0	0	0	0	0	1	0	1	0
Privacy	57	5	1	7	0	1	7	3	26	33
Security	43	2	0	2	0	0	3	0	14	14
Disability	38	5	6	16	10	4	22	10	9	29
Credit Cards	17	0	0	2	0	0	3	0	2	5
Services	45	9	14	17	7	1	24	5	26	17
Digital Signature	0	0	0	0	0	0	0	0	0	0
Email Addresses	94	87	91	92	87	80	71	72	82	85
Comment	50	33	29	36	19	12	32	20	43	31
Email Updates	33	6	14	15	6	5	10	6	14	16
Website Personalization	3	0	0	1	0	0	1	0	0	0
PDA Accessibility	0	0	0	0	0	0	16	0	2	2

Source: Author's e-government content analysis database

line services runs from 0–100 percent and is the average across each of its websites.

The privacy measure is an index that runs from 0–4 based on assessments of four different aspects of a site's privacy policy: whether the policy prohibits commercial marketing of visitor information, whether the site prohibits creation of cookies or individual profiles of visitors, whether the site prohibits sharing personal information without prior user consent, and whether the site says it can share personal information with legal authorities or law enforcement. Each of these items was coded a 0 for no and a 1 for yes. The privacy quality index was an additive scale measuring the presence of 0–4 privacy protections. Overall e-government performance is measured through rank (1 for the best performing country and 198 for the worst performning country) and numerical score (0–100 point on the e-government index described earlier in this chapter).

For the independent variables in this aggregate analysis, we used seven different measures of organizational, fiscal, and political attributes in each country. Organizational features included a measure of the extent of formal schooling.[22] This was designed to tap how well educated the labor force was in each country. In addition, we used a measure of the number of scientists and engineers working in research and development in each country per million residents from 1996 to 2000.[23] This measures the extent of technical expertise available in each nation. We also relied on an indicator that taps the extent of corruption present in each country, as measured by the World Bank.[24] This shows the efficiency with which governmental agencies function, which is related to an agency's ability to incorporate new technology on their website.

For financial resources, we employ a measure of each nation's 2001 GDP per capita in U.S. dollars.[25] This shows the extent of fiscal capacity and shows how rich or poor a particular nation is. Building on the Internet usage results of Hargittai, Kiiski and Pohjola, and Norris, our hypothesis is that countries with greater wealth will be in a stronger position to engage in technical innovatation than poorer countries.[26]

Finally, for the analysis of political determinants, we look at party competition, the extent of Internet usage, and a civil liberties scale within each country. The party competition item looks at how competitive the political parties are in each nation.[27] Internet usage is measured as a percentage of the overall population in 2001 that uses the Internet.[28] The civil liberties scale is a 1999 Freedom House measure of liberalism within each country.[29] The hypothesis is that the more liberal and democratic a country is, the greater is its ability to move toward the openness and transparency associated with electronic government.

Table 9-9 shows the regression results for indicators of the depth and breadth of online service delivery and the quality of privacy policies. The

TABLE 9-9

The Impact of Organizational, Fiscal, and Political Factors on Global Government
Online Service Delivery and Privacy Policy, 2003

	Number of Online Services	Percentage of Agencies Offering Online Services	Quality of Privacy Policies
Education	−.28 (.37)	−.07 (.10)	.00 (.00)
Corruption	−11.53 (13.35)	4.40 (3.61)	−.01 (.11)
Number of Scientists	−.01 (.01)	−.002 (.002)	−.00 (.00)
Gross Domestic Product Per Capita	.003 (.002)*	.001 (.000)	.00 (.00)
Party Competition	4.44 (11.34)	4.49 (3.07)	.10 (.10)
Civil Liberties	4.68 (5.42)	2.04 (1.47)	.03 (.04)
Internet Usage	.67 (.90)	.00 (.24)	.014 (.008)*
Constant	−14.03 (33.87)	−.34 (9.16)	−.38 (.28)
Adjusted R Square	.06	.09	.18
F	1.93	2.52*	4.55***
N	113	113	113

$*p < .05; **p < .01; ***p < .001$

Source: Author's data analysis

Note: The numbers are the unstandaradized least squares regression coefficients, with the standard error in parentheses. The number of asterisks indicates the level of statistical significance. Tolerance statistics show no multicollinearity problem in the model.

most significant predictor of the number of online services is GDP per capita. Countries that were richer tended to have more electronic services on their websites. This is in keeping with the results of other studies suggesting that economic factors are vital to policy innovation in general and e-government in particular.

There were no organizational or political factors that were important, only the level of fiscal capacity. Neither liberalism nor level of democracy were associated with e-government performance. It did not matter how competitive the party structure was or what the degree of liberalism was with respect to civil liberties. Democratic nations were no better than nondemocratic countries at innovating in regard to technology policy.

With the quality of privacy policies, however, the percentage of citizens who used the Internet was significantly related to how good the privacy policy was, not national wealth. Countries with more Internet users had more privacy protections on their government websites. This suggests that citizen usage and demand are important considerations in the public sector's effectiveness in handling privacy concerns.

On the measure of breadth of online services, none of the organizational,

TABLE 9-10

The Impact of Organizational, Fiscal, and Political Factors on Overall Global E-Government Performance, 2003

	E-Government Rank	E-Government Score
Education	−.45 (.27)	.02 (.02)
Corruption	4.90 (9.80)	−.50 (.79)
Number of Scientists	−.015 (.005)**	.001 (.000)*
Gross Domestic Product (Per Capita)	−.001 (.001)	.000 (.000)
Party Competition	−1.38 (8.32)	−.10 (.67)
Civil Liberties	1.37 (3.98)	−.01 (.32)
Internet Usage	−.14 (.66)	.05 (.05)
Constant	149.46 (24.88)***	25.46 (2.00)***
Adjusted R Square	.47	.47
F	15.41***	14.91***
N	113	113

$*p < .05; **p < .01; ***p < .001$

Source: Author's data analysis

Note: The numbers are the unstandaradized least squares regression coefficients, with the standard error in parentheses. The number of asterisks indicates the level of statistical significance. Tolerance statistics show no multicollinearity problem in the model.

financial, or political factors were significantly linked to the percentage of agencies offering online services. It did not matter how rich the country was, how competitive its political system was, how liberal the nation was, or what kind of citizen needs or demands were present. Some factor outside of the model obviously is responsible for the variation in breadth of service offerings across government agencies.

Table 9-10 presents regression results for overall e-government performance (both rank and score). In both cases, the only factor that was statistically significant was the number of scientists and engineers per million residents. The more scientists a country had, the more likely it was to score well on e-government. It did not matter whether the measure was rank or the actual score, the results were virtually identical. This measure of technical infrastructure and organizational capacity outweighed the impact of national wealth, party competitiveness, and political liberalism on electronic government.

CONCLUSION

To summarize, most governments around the world have gone no further than the billboard or partial service-delivery stages of e-government. They have made little progress at portal development, placing services online, or

incorporating interactive features onto their websites. Not only are they fail-
ing to use technology to transform the public sector, their efforts mostly con-
sist of no meaningful change or small steps forward.[30]

The factors that have constrained their progress have been fiscal resources
for online services, citizen demand for privacy policies, and number of scientists for overall performance. Similar to the U.S. state government results,
there is not a single organizational, fiscal, or political factor that limits
e-government activity. Rather, different factors are important in various areas,
depending on the particular aspect of e-government performance.

One factor that showed little tie to e-government performance is the level
of democratization. Our results show that democracies are no better at technological innovation than nondemocracies. Some have suggested that the
open and decentralized nature of the Internet will be a force for liberaliza-
tion.[31] Given the transparent qualities of IT, countries that excel at e-government
will be more open, liberal, and democratic in other system aspects, as well.
According to this reasoning, the Internet is the ultimate force for democracy
because it is difficult to control from the top of political systems.

The research reported in this chapter demonstrates, however, that there is
little correlation between democratization and e-government performance.
While IT tends to have several features that are linked to democracy, such as
openness and transparency, this study also shows that democracies have attributes that slow down policy innovation, such as competing centers of influence, difficulty in top-down control, and lack of unified political power.

In order to make progress on e-government, public officials must overcome
bureaucratic intransigence and centripetal forces that slow innovation and
prevent the public sector from moving forward rapidly. Nondemocratic systems appear as adept as democracies at overcoming political barriers. Therefore, it is little surprise that we see a number of authoritarian countries performing quite well on e-government. Indeed, since the dominant vision of
e-government emphasizes service delivery, not democratic system transformation, it helps to explain why nondemocratic countries have demonstrated
great success at putting services online.

Regardless of the type of political system, in looking toward the future, it is
important that nations create government portals that serve as gateways to a
particular country's websites and offer a one-stop web address for online services. Some countries have adopted portals and put services for citizens, businesses, and government agencies in one place. This is a tremendous help to
citizens interested in making use of online resources. Portals are helpful from
the citizen standpoint because they offer more uniform, integrated, and standardized navigational features. One of the weaknesses of many national websites has been the inconsistency of their design features. Government agencies guard their autonomy very carefully, and it has taken a while to get
agencies to work together to make the tasks of citizens easier to undertake.

Common navigational systems help the average citizen make use of the wealth of material that is online.

Governments need to find ways to take advantage of features that enhance public accountability. Simple tools such as website search engines are important because such technologies give citizens the power to find the information they want on a particular site. Right now, only one-third of government websites are searchable, which limits the ability of ordinary citizens to find information that is relevant to them. This limits citizen ability to use and manipulate information.

The same logic applies with regard to features that allow citizens to post comments or otherwise provide feedback about a government agency. Citizens bring diverse perspectives and experiences to e-government, and agencies benefit from citizen suggestions, complaints, and feedback. Even a simple feature such as a comment form empowers citizens and gives them an opportunity to voice their opinion about government services they would like to see.

Countries need to devote more effort to update their sites on a regular basis. Some appeared as if they had not been updated in several years, with the result being that information on the Web is seriously outdated. If countries both update and place more material online, it would encourage citizens to make greater use of e-government resources. In the current situation, citizens have few incentives to go online when government sites are not updated, contain inaccurate information, or have broken links or email contact information. Only by maintaining these sites and repairing broken features will citizens and members of the business community see online government as a help to themselves.

The issue of how to pay for portals and other e-government costs remains a pressing challenge for almost every country. The start-up costs of e-government are extensive and small or poor countries have difficulty reaching the economies of scale necessary to pay for the technology. While a few sites employ commercial advertising or user fees for their public sector sites right now, there still are risks either in commercializing e-government or relying on user fees directed at website visitors. The former creates potential conflicts of interest for government agencies if their websites become dependent on commercial revenue. The latter disenfranchises people of more limited means and widens the digital divide between rich and poor.

Clearly, a major problem of e-government is the up-front costs of developing a website and putting information and services online. Right now, many nations appear to be undertaking these tasks in isolation from other nations, thereby robbing each country of the opportunity to achieve economies of scale that would lower the per-unit cost of official government websites. Smaller and poorer countries should undertake regional e-government alliances that would allow them to pool resources and gain greater efficiency at building their infrastructure. Such collective efforts give citizens interested in a region one place to find information that cuts across individual nations. At the same time, such a site also offers economies of scale to specific countries in placing cultural and religious material on the Internet. These efforts at regional cooperation are valuable because they put countries in a position where they can share knowledge and expertise as well as lower their overall budget costs.

CHAPTER 10

Democratization and Technological Change

WHEN THE WORLD WIDE WEB initially appeared on the technological scene, proponents immediately heralded it as a revolutionary device.[1] The Web had a number of features that appeared quite advantageous from the standpoint of societal transformation. The Internet decentralized communications, was nonhierarchical, allowed people who were geographically remote to communicate asynchronously, was convenient due to its 24/7 availability, had two-way communication capabilities, and was interactive.[2] In short, it contained so many obvious advantages in the eyes of its developers that there was no doubt it would radically alter society, politics, and commerce.

Writers quickly seized on these features to argue that the Internet would usher in a new era that would transform government performance and democracy itself. Citizens would communicate quickly and easily with public officials. Economies of scale would allow technology to improve service delivery in the public sector. Bureaucrats would become more responsive to the concerns of the citizenry. Public trust would be restored because government would operate in an effective, efficient, and responsive manner. Direct democracy and citizen participation in elections would be facilitated because the costs of information acquisition and political communications would drop to nearly zero.

In reality, though, many factors have constrained the ability of the Internet to alter the manner in which government and democracy function.[3] Public sector performance has improved more slowly than early advocates predicted due to the costs of technology, bureaucratic fragmentation, group competition, and poor political leadership. In terms of the four stages of e-government (billboards, partial service delivery, portals with fully integrated services, and interactive democracy), most U.S. sites are somewhere between stages two and three. And when one looks around the world, most non-U.S. government sites remain mired in stages one or two. Not only has there been inconsistency in the pace and breadth of technological diffusion throughout the public sector, a digital divide has limited Internet access by those who are poor, not highly literate, or disabled.[4]

In this chapter, I step back from these findings and discuss the link between democratization and e-government performance. After discussing democratization and the limits on technological change, I compare the Internet to earlier technological advances, such as the printing press, telegraph, tele-

phone, radio, and television. I argue that the slow rate of change found with
e-government is not atypical. With most new creations, there were organizational, fiscal, and political attributes that slowed diffusion and limited the
transformational potential of the technology. Rather than being unusual, the
incremental nature of Internet change is consistent with the history of many
past technologies.[5]

DEMOCRATIZATION AND E-GOVERNMENT PERFORMANCE

One important result of this study is the weak tie between democratization
and e-government performance. As discussed in chapter 9, nondemocratic
governments are as likely as democratic governments to rank highly on
e-government performance. In terms of overall quality, authoritarian nations
such as Singapore, Malaysia, and China have performed well and in some
cases rated more highly than those of several European democratic systems.

In addition, nondemocratic lands such as Singapore and Hong Kong
scored well on privacy considerations, as measured by the percentage of their
websites displaying a privacy policy and the four-point privacy quality index.
For the latter index, which looked at the extent to which four important indicators of an effective privacy policy were present on the site, Singapore rated
third (following Canada and Australia) and Hong Kong rated twentieth out of
the 198 nations around the world.

The limited impact of democratization is demonstrated in aggregate models seeking to explain global e-government performance. Political factors
such as party competition and civil liberties ranking had no association with
overall e-government performance. Economic considerations were more important than institutional factors linked to democratic systems in how well a
country integrated technology into the public sector.

These results are noteworthy because the openness and transparency associated with the Internet would appear to be strongly linked with democratic
regimes. At least in the abstract, democracies have advantages in terms of
e-government performance since they are open and transparent, and digital
delivery systems encourage officials to move in that same direction. E-government is not about requiring political connections for service delivery or re-
stricting public information, but placing services and information online.

Indeed, many early Internet proponents made bold claims that digital technology would revolutionize politics, weaken authoritarian regimes, and empower ordinary citizens. Individuals such as Reed Hundt, the former Federal
Communications Commission chairman, bluntly stated that "The Internet
Changes Everything."[6] Other technology buffs have enthused that dictatorships would have difficulty controlling information in a decentralized Inter-

net system and that citizens would gain greater power over formal government decision making.[7]

As shown in chapter 2, however, the factors that facilitate e-government development are numerous. There are a variety of organizational, fiscal, and political features that encourage the incorporation of technology into the public sector. While some of these factors are associated with democracies, many of them are not.

Some authoritarian countries have been successful in implementing e-government because they have top-down political structures, compliant private sectors, and a unified policy vision. As noted in earlier chapters, a major barrier to e-government development has been difficulty overcoming bureaucratic and political intransigence. In democratic systems, it is challenging to get different agencies to work together or to convince career administrators to try new approaches.

This is less of a problem in authoritarian regimes because officials have the power to overcome political and organizational barriers to technological innovation. Through their hierarchical structures and ability to issue firm edicts, nondemocratic systems are well positioned to provide a clear vision for digital government and to carry out that vision. Once political elites in these regimes have decided something is an important priority, they face fewer obstacles than in democratic systems in bringing that vision to fruition.

Democracies often have multiple veto points and numerous means for contesting policy actions. Merely because a leader wants something to happen does not guarantee that action definitely will take place. Groups can fight proposals they feel are harmful and use both public and private communications channels to stymie initiatives. In democratic systems, this openness often has slowed progress in the digital area and complicated the integration of technology into the public sector.

Several authoritarian countries have proven quite adept at building political coalitions and assembling financial resources for digital government. They have overcome factionalization and political intransigence. Leaders have gotten competing agencies to work together to take advantage of new technology. Indeed, technology has become a way for these nations to demonstrate their scientific progress and success at serving businesses and citizens.

BUILDING SUPPORT FOR TECHNOLOGY INITIATIVES

This research also has ramifications for how political leaders build support for technology initiatives. In its early days, e-government started as a nonpartisan, technocratic reform that would improve public sector performance and democracy. Government would become more efficient and effective, and agencies would be more responsive to ordinary citizens. By taking advantage of the interactive features of the Internet, digital government would narrow the gap between citizens and leaders, and make people feel more positive about the public sector.

In making these arguments, e-government fell within the issue type known as a constituent (or self-regulative) policy area. Developed by Theodore Lowi, this typology distinguished four types of policy areas: redistributive, distributive, regulatory, and constituent.[8] Depending on how concentrated the benefits are (broadly or narrowly) and the intensity of the conflict (zero sum versus non–zero sum), public policies vary in the particular politics that emerge.

Constituent policies are non-zero-sum arenas where benefits affect a narrow pool of citizens. They are marked by low visibility, expert domination, nonpartisanship, and the absence of political controversy. They represent a technocratic domain where politics is muted and partisanship generally avoided. Self-regulation is the norm as these areas attract little attention from the media or the general public, and do not engage large numbers of people. Experts largely determine the course of public policy, and professional administrators implement their ideas.

In contrast, some policy areas are redistributive in nature. They take from one segment of society and transfer those resources to another. Large numbers of providers and recipients are involved (such as in the case of welfare), and the politics tend to be highly partisan and controversial. Citizens pay close attention to redistributive policies because the conflict is zero sum and affects broad groups of people.[9]

Regulative policies such as environmental legislation and consumer safety protections are zero sum, but involve more concentrated benefits. This lowers the visibility of the area and limits political conflict to those interest groups involved in the particular area. The general public is not engaged in this area, but groups whose economic interests are affected are very involved in the process. They attempt to influence bureaucratic rule making and turn regulations to their benefit. There is some partisanship involved with these efforts, but not to the extent of that in redistributive areas.

Distributive policies are again non–zero sum arenas and involve broad categories of beneficiaries. Legislators tend to logroll in this area, and agree to support each other's projects. This reduces the overall level of conflict and undermines the potential for partisanship. Highway bills and federal grants are the classic example of this policy type. Large amounts of money are spent, but the fact that many geographic areas benefit limits the intensity of political conflict.

For much of its history, e-government has fit within the area of low conflict, low partisanship, limited visibility, and narrow benefits represented by constituent policies. Conflict has been muted because digital policy is a technocratic area dominated by experts. There has not been much partisan controversy because e-government is seen by the public in positive terms. There is not a Republican or Democratic way of building government websites. Less than a third of the population are heavy users of the Internet so beneficiaries of digital government represent a relatively small group of people.

In order to build political coalitions on a topic of low visibility and narrow benefits, proponents relied on technocratic appeals and the promise of improved public sector efficiency, effectiveness, and responsiveness. According to its proponents, cost savings would come as more people started to use the Internet and visited government websites. Benefits would flow broadly to large groups of recipients, and citizens as a whole would feel more attached to government agencies.

Digital government is moving towards the characteristics of other policy types, however. It is becoming more partisan and more controversial. Costs are a rising concern as local, state, and national governments face budget deficits and have to balance IT spending with that for health, education, welfare, and defense. As Internet usage levels rise, it is shifting from narrow to more broadly defined recipients. This engages a larger number of individuals and organizations, and attracts more press coverage.

As these characteristics unfold, both the politics of the situation and the manner in which technology is integrated into the public sector are affected. Partisan differences among citizens are emerging that show how different parts of the general public feel about digital government. As shown in chapter 7, Democrats are more likely to be favorable about e-government than are Republicans. This is in keeping with the generally progovernment sentiments of Democrats. Even after controlling for factors such as education, race, age, and gender, there is a political divide in the United States in the sympathy people have toward government websites.

There also is a change in attitudes among unions that is starting to affect e-government. Unions are organizing to make sure e-government does not endanger worker rights or health. The American Federation of Teachers Public Employees (AFT) organization undertook a study of digital government and technological change that examined the impact on teachers.[10] In its report on that study, the AFT noted that "public employees are facing new health and safety challenges with the introduction of new systems of technology in the workplace."[11] Among the chief concerns of union members were carpal tunnel injuries and new work schedules.

According to the Peter Hart Research survey in June 2002 sponsored by AFT, 26 percent of AFT public employees have developed health problems that arose through the use of computer or other kinds of technical equipment.[12] In addition, 17 percent of AFT public employees felt that employers "excessively monitor employee use of e-mail and the Internet."[13] Based on these concerns, the AFT report suggested unions establish health and safety committees to monitor the impact of new technology, develop joint labor/

management committees, clarify membership rights in regard to the use of email and the Internet, and create new employee training programs.

As more money goes into e-government, the media are providing more coverage and some of this reporting is uncovering cost overruns and procurement problems. As discussed in chapter 8, journalists are applying investigative tools to digital government and finding that government officials are rewarding friends and allies with contracts, and that new technology is not performing as well as it should. The resulting news stories are narrowing the distinction between traditional and e-government, and leading some observers to feel that online government may be little better than its bricks-and-mortar counterpart.

Political partisanship, more aggressive union organizing, and more critical news coverage undermine the technocratic vision of e-government that has sustained this domain since its inception. Both unions and management see e-government as a new policy development to be contested. As IT becomes the object of political and partisan conflict, the bipartisan coalition that characterized its early phase is becoming more tenuous. E-government is emerging as a political football in the manner of many issues from education to health care. In 2002, for example, digital government was a contentious issue in several gubernatorial campaigns around the country (see chapter 2).

If this trend continues, e-government will move from a "constituent" or self-regulative policy area to either a regulative or redistributive issue, depending on the scope of beneficiaries.[14] Drawing on the analogy of the early days of radio, Eszter Hargittai predicts that government regulation of the Internet is "inevitable" given the power of digital technology.[15] Conflict over technology policy is becoming more zero sum and it will require visible and partisan appeals to build supporting coalitions. No longer will technocratic appeals sustain the coalition. Instead, supporters will have to justify IT expenditures based either on a conservative message (such as cost savings and improved efficiency) or a liberal message (closing the digital divide or providing more universal access).

If technology is justified relying on a conservative perspective, then leaders will have to explain how e-government is saving money and improving the overall efficiency of government. Studies will be required that show economies of scale through technology. Proponents will need to show that as more people access digital resources, it is possible to cut back on bricks-and-mortar government. That is the major way in which online services will allow officials to save money and become more efficient.

Alternatively, new spending can be justified on more liberal grounds that technology promotes more universal access and closes the digital divide. This is a very different appeal than outlined above, but one that could entice citizens to want to invest in new technology. Digital government could be sold as a tool for citizen empowerment and a way to involve ordinary folks in government policymaking. Similar to other direct democracy initiatives such as primaries and referenda, online government could be used to promote a new vision of online democracy.

Although either one of these appeals could be effective at mobilizing public support for technology, each fundamentally alters the political dynamics of the issue area and complicates the task of incorporating new technology into the public sector. These approaches would politicize what previously has been a technocratic policy area dominated by experts. They would require politicians to focus on particular constituencies and use ideological arguments to justify technology spending. This inherently is more partisan and conflictive than the technocratic arguments used in the early days of digital government.

The transition from technocratic reform dominated by nonpartisan experts to an ideological area that either serves business interests or is used to empower the general public would make it much more difficult to assemble political and financial resources for digital government. Similar to areas such as education and health care, progress would depend more on specific coalitions than general support for technology. It would be impossible to assemble the broad-ranging, nonpartisan coalitions that have supported technology in the past. Rather, new initiatives would have to be justified based on the concrete results produced for particular beneficiaries.

LIMITS ON TECHNOLOGICAL CHANGE

With political and economic pressures impinging on e-government, the power of the Internet and other digital technologies to transform the public sector has been limited. Both in the United States and around the world, tight fiscal resources are constraining the ability of Internet technology to transform the public sector and democracy itself. Most U.S. governmental units are putting a rather small percentage (1–2 percent) of their overall budget into IT. Other nations, by and large, are devoting even less than this to digital government.

Right now, there is little evidence that e-government is saving money because the public sector is having to maintain the old bricks-and-mortar delivery system while also developing digital models. Part of this "two systems" problem is that relatively low numbers of people are visiting government websites and accessing online services. The digital divide in IT forces officials to maintain multiple sets of service-delivery systems: traditional service delivery through telephone calls, mail requests, and in-person visits as well as electronic service delivery. Having parallel processes for delivering information and services limits the ability of governments to achieve the cost savings that proponents envisioned. True cost savings will come only when online usage grows to the point where the digital divide closes and officials can trim the costs of traditional government.

It is not just financial issues that limit the use of new technology. Organizational settings and political dynamics constrain the rate of technological change.[16] Despite a bureaucracy that is generally amenable to Internet technology, government agencies are slow to innovate because technology requires shifts in job titles, division of labor, job definitions, and hours of operation. The benefits and costs of technical enhancements must be very clear for the public sector to make sweeping changes in how the agency functions. This is especially true for agencies that are not externally directed, such as the courts.

Interest groups are fighting over government contracts and whether electronic services such as tax filing should be handled by the private sector through commercial filing services or by the public sector through free on-line filing software. As demonstrated in chapter 5, private tax preparation companies pushed to keep the IRS from developing free, online tax filing that would compete with its own commercial products.[17] Even though a number of states successfully have implemented Internet filing, industry sources fought effectively to keep the IRS from using available technology to develop such a capability. These groups were victorious when the Bush Administration sided with them and agreed to a three-year moratorium on direct federal online filing. Rather than build a governmental tax-filing system, the administration is relying on a private system run by commercial tax preparers, with free service offered to low-income filers.[18]

These organizational, fiscal, and political factors have tremendous consequences for technological change. As shown in table 10-1, both individual adoption and institutional change are required for there to be widespread social and political ramifications of new technology. Individual adoption decisions are affected by one's ability to pay for new inventions, how favorable someone feels toward technology, and the benefits that technology offers for day-to-day living. In general, Americans are very favorable to technology and the Internet. Indeed, technology is considered one of the country's greatest achievements in the twentieth century. People love having the latest consumer gadgets and using them to make themselves more efficient.

For there to be widespread social and political ramifications, however, both individual and institutional forces must be moving in the same direction. Individuals must see advantages in terms of adopting technology and government institutions must have the resources, organizational incentives, and leadership to transform the public sector. To this point, this kind of epochal change has not happened. The constraints noted in this book have limited the ability of government officials to embrace technology and use it to reorient how they perform their agency functions.

Even though use of the Internet is popular as an individual activity, most

TABLE 10-1
A General Model of Technological Change

Individual Adoption
↓
Institutional Change
↓
Widespread Social and Political Ramifications

Source: Author compilation

Americans employ it for nonpolitical and nongovernmental activities. As shown on table 10-2, the most popular online activity in 2001 was emailing (named by 84 percent of users aged 15 or older), followed by product searches (67 percent), news, weather, and sports (62 percent), and playing games (42 percent). Accessing government services was the only item on the list related to e-government, and it was rated seventh out of fifteen activities. Thirty-one percent of users cited that as a favorite Internet activity.

These findings are consistent with the political disengagement that most Americans feel toward government. Ordinary people are more focused on using the Internet to better their personal lives, not the public sector. This

TABLE 10-2
Most Popular Online Activities, by Percentage, 2001

Email	84
Product Search	67
News, Weather, and Sports	62
Playing Games	42
Product Purchases	39
Health Information	35
Government Services	31
Help on School Assignments	25
Viewing TV, Movies, or Radio	19
Online Banking	18
Chat-Rooms	17
Job Searches	16
Stock Trading	9
Making Telephone Calls	5
Online Education Courses	3

Source: 2001 U.S. Census Bureau Current Population Survey

limits the ability of policymakers to employ the Internet as a tool for civic engagement and democratic revitalization.

These nonpolitical preferences help to explain why many government officials have followed a vision based on service delivery as opposed to system transformation. Rather than seeing the Internet as a tool for citizen empowerment and public responsiveness, they have put more money into information and services than accountability-enhancing and interactive features that strengthen the role of the general public.[19] Government websites have not incorporated many interactive technologies, reassured the public about security and privacy, or promoted access by special-needs populations. Change from year to year has been slow, and most government agencies have encountered a number of barriers that have limited their ability to produce larger-scale change.

The major uncertainty of the current era is whether the slow progress seen so far will lead to incremental change that is not significant or will add up to the "slow but steady" secular change that is quite significant over a period of time. For example, it took the automobile decades before individual ownership and usage spread among the general population, facilitated the flight of higher income people from the central cities, and produced major social and political change. The long-term secular change proved quite revolutionary even if it took a considerable time before the car's social and political ramifications unfolded.

It is hard to provide a definitive answer regarding incremental versus secular change, but preliminary evidence suggests some optimism in terms of how the public views innovation through the Internet. Despite several concerns in the area of privacy and security of online transactions, the public remains quite favorable about Internet delivery systems. People like the technology and see it as one of the most beneficial improvements in social life of the last several decades. This generally favorable view about Internet technology bodes well for e-government.[20] As long as the public sector is able to keep adding services without requiring a lot of financial resources, public interest in and usage of e-government will continue to rise.

The major things e-government must avoid are partisan scandals or contracting abuses that would undermine public favorability toward the Internet. If there are problems in out sourcing e-government service delivery to private companies or governments are not able to maintain the privacy and security of online transactions, public support for e-government will drop and people will view electronic governance as no better than traditional governance.

Obviously, researchers should not rush into definitive judgments regarding the ability of e-government to transform service, public sector performance, democratic responsiveness, or citizen trust in government. There are around 87,000 government units in the United States alone, plus thousands more for the 198 nations around the world. It will take a long time to determine whether e-government becomes a vehicle for the revitalization of public sector performance and democracy.[21] Many factors constrain the ability of government bureaucracies to remake themselves.

COMPARISONS TO THE EARLIER TECHNOLOGICAL ADVANCES

In falling within a model of limited change, the Internet is not unusual. A review of past technologies shows that a slow rate of transformation is typical. Many of the organizational, fiscal, and political features that constrain change in the contemporary period have slowed the diffusion of earlier technologies.[22] Historical cases of new technology from the printing press and the telegraph to the telephone, radio, and television were considered earth-shattering when first introduced. As with the Internet, however, proponents oversold the revolution. Sometimes, it took decades or even a century for the invention's full potential to be embraced by individuals and implemented by government organizations.

For example, the printing press was said to undermine the power of traditional authorities and enable new centers to emerge. The invention of movable-type printing greatly simplified the dissemination of information. By standardizing the production of books, which previously had been hand-copied, the printing press made it possible for publication costs to be reduced, and book usage thus expanded. Yet it required an extraordinary amount of time for the technology to bring about the cultural, political, and social consequences now ascribed to the printing revolution. As an illustration, it took a century for the printing press to diffuse the eight hundred miles from Germany to Russia.[23] For the press to achieve significant impact in the 1500s and 1600s, there had to be an increase in the number of books being published and an expansion in the number of publishers who could disseminate knowledge and information.[24]

Once these intermediate steps were reached, traditional social authorities such as feudal lords and the Catholic Church lost their ability to control information. Power became more diffused and new centers developed. Through the creation of "canonical" texts—meaning books that were faithful to the original—scholarship improved and a Renaissance in human understanding blossomed.[25] This became the basis of the scientific revolution and the formation of the nation-state.

More recently, the telegraph was said to transform patterns of news gathering because of the ability of correspondents at the front lines to send updates instantaneously to their home offices. In 1844, the first telegraph message, "What hath God wrought?" was sent from Washington, D.C., to Baltimore, Maryland. Using wires that ran the forty miles between the two cities, inventor Samuel Morse unleashed a technological change that speeded up cross-country communication. Rather than relying on the mail and information sent by personal messengers, the telegraph created the capacity for near-instantaneous information delivery. Speaking of this invention, *New York Herald* editor James Gordon Bennett noted that because the telegraph "communicates with the rapidity of lightning . . . the whole nation is thus impressed with the same idea at the same moment."[26]

Soon, newspapers all across the country were relying on the telegraph to transmit news bulletins from field correspondents to the home office. The Mexican-American War of 1846 allowed newspapers for the first time to use the telegraph to get the latest developments on the conflict and run stories informing readers about important news. Yet it took several decades for this new communications device to become widespread. It was not until 1880, three decades after its invention, that a bare majority of morning papers (52 percent) were using the telegraph to provide the latest material from regions far afield.[27]

In 1876, Alexander Graham Bell created a new device called the "telephone" that sent spoken words over electrical wires. Employing lines that connected different places, Bell's invention allowed people far apart to have real-time conversations that both parties could hear. While the first conversation from Bell to his assistant—"Come here, Watson. I want you"—was quite brief, it paved the way for what contemporaries called the "speaking telegraph."[28]

The long-term impact of this invention was quite substantial. The telephone started the process that emancipated people from communications barriers based on geographic distance. It created more freedom and convenience for those who wanted to talk with other people who were miles apart. By providing audio links across disparate points, the telephone narrowed the social and political isolation caused by distance and speeded up the process of communication across the United States.

It took years, however, for use of this new gadget to become widespread among the American population. In the first month following the invention of the telephone, only six phones were sold. By 1900, twenty-four years after its creation, just 1 percent of American households had a telephone. At that point, the telephone was very much an elite luxury enjoyed by wealthy and privileged members of society. A decade later, the proportion of Americans owning a telephone had risen to 10 percent. It was not until 1920, when 36 percent had a telephone, and 1930, when ownership increased to 41 percent of the U.S. population, that telephone access became more widespread.[29]

The slow pace of adoption of the telephone sharply limited its ability to transform communications. Family life and political affairs could not be affected until large numbers of individuals gained access to the device. The technology was useful only when someone a person wanted to call also had access to a telephone.

In the 1920s, radio emerged as a vehicle for news and entertainment when

Pittsburgh station KDKA became the nation's first commercial radio broadcaster.[30] The station provided up-to-date coverage of the 1920 presidential campaign for its listeners. Within three years, six hundred radio stations were broadcasting live news reports over the airwaves, and this wireless technology was one of the most popular ways of hearing the news, music, and entertainment.

Radio had several qualities missing from other media of the day: intimacy, directness, and the potential for new updates throughout the day. By bringing audio sounds right into people's homes without any extra wiring, radio provided a personal connection that was not present with newspapers (even with multiple editions each day) or printed circulars. The audio connection was direct and spontaneous and people appreciated the close contact provided by this technology. As new developments emerged during the course of a day, that information could be delivered straight to listeners around the country. News dissemination was not dependent on print schedules or newspaper production. With the click of a radio dial, anyone could hear a variety of news and entertainment options.

As a sign of the more rapid diffusion of technical innovation in the twentieth century, the expansion in ownership of radio occurred much more quickly than with the telephone. In 1922, only 1 percent of American households had a radio, but this number rose to 46 percent in 1930, and 82 percent in 1940.[31] This was a much more rapid increase than had occurred with telephones just a few decades before. With that invention, it took 54 years after its invention to reach an ownership level of 41 percent. With radio, the 46 percent figure was reached merely a decade after its development.

The rapid dissemination of radio technology and its relatively low cost for consumers facilitated its ability to alter the political landscape. During World War II, correspondents such as William Shirer and Edward R. Murrow became famous simply through their ability to bring radio news into the kitchens and living rooms of many Americans. Politicians such as President Franklin Delano Roosevelt used fireside radio chats to personalize their official announcements and instill hope in a discouraged American public. Families all across the United States sat around the radio hoping to catch the latest information on how the war was going, and what might be happening to loved ones away in battle.

The year 1946 saw the first coast-to-coast broadcasting of television. This new technology joined the immediacy of radio with the visual technology of moving pictures. Television had a number of virtues that made it immediately popular. Most important, of course, was the visual aspect of being able to see images broadcast from far away. As opposed to being able to describe what was happening, television allowed news and entertainment producers to show what was going on. This "you are there" feature appealed to viewers and led to a rapid expansion of the television industry.

Although only 9 percent of American households had a television in 1950, this number skyrocketed to 88 percent in 1960. This sharp jump in access to television demonstrates how much more quickly technical inventions were diffusing throughout the population by the mid-twentieth century. In contrast to the telephone, which took 60–70 years to penetrate into 80 percent of the country, and radio, which took two decades, television ownership rose from 9 to 88 percent in the space of one decade. This near-saturation adoption of television occurred just 14 years after the initial experiment in coast-to-coast broadcasting.

Not surprisingly, given the rapid diffusion of television among the American public, this invention quickly became a political force in elections and governing. With the advent of the first televised presidential debates in 1960, television was commonly thought to have been decisive in shaping how people saw the major political candidates.[32] Challenger John F. Kennedy was young and telegenic, and staged rallies and photo opportunities designed to convey the message that he would get the country moving again. After the staid decade of the 1950s, Kennedy's good looks and personal charisma projected well on television and fit nicely into what ultimately became a winning campaign strategy.

Kennedy's electoral victory that year was decisive in leading observers to conclude that a new "television" era was emerging in American politics. During this epoch, it was said that politicians would use television to communicate visually with voters. The rise of telegenic politicians who could use this new medium to talk with the public allowed those individuals to gain popular support in the process. Journalists soon were glamorizing the Kennedy presidency and showering extensive attention on politicians who looked and sounded good on television.[33]

Academics, though, cast doubt on these interpretations. Rather than glorifying media effects, public opinion researchers argued that longer-term social and political forces were more important to voting behavior.[34] Viewers were not idiots who believed everything television reporters told them. Instead, they filtered incoming beliefs through the lens of partisanship and ideology. This limited the ability of television to be politically influential.

With this kind of uneven history, technology often has emerged in surprising ways. The development of electronic-delivery systems in the public sector took a long time after ARPANET initially appeared in 1969. The technology diffused slowly at the beginning, but then did so more quickly following the creation of the World Wide Web in 1991. Since that time, public usage in the United States has increased from 4 percent to 49 percent in 2001.[35] Current estimates indicate that around 60 percent of the American population employ the Internet.[36]

The recent expansion of Internet usage at the individual level encourages some observers to conclude that technology unfolds in a positive and transformational manner for politics and society as a whole.[37] As has been emphasized throughout this book, though, this view is not entirely correct. Much as has happened with past inventions, technology can diffuse widely throughout the individual consumer market and still not transform social, economic, and political institutions. Instituttutional change tends to be far slower than individual behavior because the former is more subject to organizational, fiscal, and political dynamics.

In addition, not all technological change moves in a positive direction. Change often is multifaceted. It looks more like a checkerboard and produces consequences of wildly different character.[38] Technology can unleash both positive and negative ramifications, and be quite complex in its overall consequences for society, politics, and democracy. In its early days, the convenient character of computers and their capacity for two-way communication were seen as a boon to democratic participation and representation.[39] The reality, though, has been less radical, and personal computers have not been a panacea for society or the political system. Rather, they bring a variety of risks and hazards, as evidenced by spam, worms, viruses, and hackers, and not all Americans have access to Internet technology.

In some respects, e-government has improved productivity and accelerated communications. If this persists, then the Internet may make the public feel better about government. Public cynicism about traditional government has remained strong over three decades of scandals, inefficient performance, and poor economies.[40] It will take major improvements in e-government performance and evidence that technology is responsible for the improvement in order for the public to transform itself into trusting, participatory, and noncynical citizens. As shown by the questionnaire priming experiment, there is some evidence that detailed exposure to the issue of e-government leads to greater confidence in the public sector.

It is unlikely, however, that the Internet is going to make rapid progress in involving people in direct democracy.[41] Most policymakers do not see the Internet as a tool for citizen empowerment. They are not employing electronic technology to cast votes online and draw citizens into civic discourse or policy deliberation. The vision that is most prevalent in the public sector is not sympathetic to grass-roots democracy. Rather, the vision rests much more on maintaining representative democracy, with leaders acting on behalf of their constituents. By embracing that orientation as opposed to social and political transformation, officials are watering down the most radical features of the Internet.

A national opinion survey shows that this aversion to some types of public empowerment is shared by many citizens. According to a 2000 Council for Excellence in Government national poll undertaken by Peter Hart and Robert Teeter, 59 percent of people oppose using the Internet for online voting.[42] This opposition has remained high in following years, as well. In 2003, for example, a follow-up Council survey found that 67 percent were against online voting.[43]

Paramount among citizen worries in both surveys were the privacy and security of using the Internet in this way, and concerns over whether online voting would be associated with voter fraud. The American military has cancelled plans for online voting due to security concerns.[44] In addition, the public is not persuaded the Internet should be used for direct democracy or public deliberation on legislative proposals. In keeping with America's emphasis on representative rather than direct democracy, many people are skeptical of the Internet's utility for empowering ordinary folks for collective political action.

Although interactive features are available that could draw citizens further into actual decision making, most government units are not incorporating those features into their websites. With few exceptions, almost no American city or state is using the Internet to improve voter registration rates or levels of voter turnout in the elections process. They are not using the two-way communication power of the Internet to bring citizens into city council meetings or legislative debates. They are not focusing on welfare recipients, the poor, or the disabled, but on middle class and business audiences. The old cable television experiment with direct democracy in the form of two-way QUBE technology remains a distant memory for most government planners.[45]

THE FUTURE OF E-GOVERNMENT

In looking beyond the current situation, it is apparent that significant changes are needed to improve the ability of government planners to harness the transforming power of the Internet. Many of the factors that limit e-government have much more to do with organizations, financing, and political dynamics than with technology per se. The technology to improve democracy and address the needs of special populations already is available. Rather, it is a question of organizational and political will to take full advantage of the benefits of the Internet.[46]

The first change that would make a difference is the streamlining of government technology offerings. Greater progress needs to be made in creating websites that have uniform, integrated, and standardized navigational features. Right now, government websites have a disjointed quality that impedes communication and citizen usage across sites. Every time a citizen logs on to a new government site, he or she must learn how that particular site is organized and where to find particular kinds of material. More consistency across e-government sites would make it easier for citizens to use online materials.

Second, there needs to be greater cooperation on the part of government agencies such that one-stop portals and cross-agency offerings are integrated. Government agencies guard their autonomy very carefully, and one of the biggest barriers to e-government improvement has been getting agencies to work together to make sites user-friendly. Portal sites that integrate information regardless of the agency source are convenient for citizens and help them avoid the problem of not knowing where to find particular services or information.

Third, agencies need to publicize the existence of government service portals. According to a 2000 national survey conducted by Peter Hart and Robert Teeter for the Council for Excellence in Government, anywhere from a third to a half of Americans (depending on the level of government) have logged onto a public sector website. While some of this access problem reflects lack of availability to computers and the Internet, many citizens clearly need to be educated as to the existence of online services and information. Marketing tools such as placing the portal address on state documents, putting the address on vehicle license places, and using televised public service announcements would help the average citizen learn how to access e-government resources.

Fourth, governments need to appoint a high-level administrator in charge of electronic governance.[47] Most states have a chief information officer who performs this task, although the person often does not have independent budget resources or clout within the administration. In the U.S. federal government, despite legislative requests, while individual agencies each have a chief information officer, but until the passage of the E-Government Act of 2002, there was no high-ranking person charged with overseeing the development and execution of e-government. This has limited the ability of electronic governance to move forward.

While we have uncovered little evidence of transformational change in the e-government area, there is the possibility of more extensive change emanating from the Internet in the longer term.[48] There are several factors that presage an optimistic future for e-government: the engagement of young people in Internet technology, societal trends that encourage the adoption of new technology, and the generally favorable views that most Americans hold toward Internet technology.

Young people remain the age group most committed to and enthusiastic about Internet technology. National surveys show that older citizens are much less likely to visit government websites or use online services. Those who are younger are the most likely to take advantage of electronic options. As these young people age, they will demand electronic governance and want options that allow them to access public sources online. They will not want to wait in line but will prefer going online to receive information and services. In the longer run, generational changes will speed the ability of the Internet to transform government agencies.

In addition, societal trends favor the spread of new technology. As e-commerce and e-trading become more widespread, people will expect similar innovation from the public sector. The rise of interactive features on commercial sites will put pressure on government agencies to add similar features. The ease of use associated with purchasing private products online will lead people to expect the same convenience in their dealings with the public sector.

Finally, the positive impression that people have about e-government bodes well for its future. Among the various inventions that have appeared over the past twenty years, Americans consistently have rated the Internet as one of the most beneficial. It is seen as productivity-enhancing, service-improving, and cost-reducing. While there are clear problems that concern Internet users, such as those of privacy and security, the upside of Internet technology clearly outweighs the downside in the eyes of most people. This suggests that the future of electronic governance remains quite bright as people become more familiar with and comfortable about online service delivery.

The digital divide, however, continues to limit progress on e-government.[49] About one-third of Americans and many more people around the world have little access to the Internet. These individuals fall outside the digital world and have few prospects for getting connected electronically. The large number of citizens who do not have access to the Internet harms e-government at the economic, social, and political levels.

In terms of economic consequences, it is difficult to achieve the cost savings envisioned by proponents until the per-unit costs of digital technology fall appreciably. For government planners to solve the two systems problem of parallel service delivery, more and more Americans need to access services online. If usage levels rise, it becomes possible to reduce the paper flow through more traditional service-delivery avenues (in-person, telephone, or mail) and cut the costs of government.

At the social level, the large number of people without adequate access to the Internet intensifies equity problems. As long as some individuals, generally richer and more highly educated citizens, have access to technology, they gain advantages through access to government information and service delivery that are not available to others. This exacerbates the gap between the information haves and have-nots, and reinforces existing class divisions within the country. According to Ben Shneiderman, universal usability is defined as "having more than 90% of all households as successful users of information and communications services at least once a week."[50] Right now, the United States has an Internet penetration rate of around 60 percent, so it falls far short of universal accessibility.

Political progress has been slowed because tools for technological empowerment have not been incorporated into the vision of e-government. Rather than using technology to improve responsiveness and enhance accountability, government officials have focused more on delivering services to businesses and middle-class users. This undermines the transformational potential of the Internet and restricts the ability of technology to improve the functioning of democracy.

To deal with these problems, the United States enacted the E-Government Act of 2002. It created an office of Electronic Government within the Office of Management and Budget. The legislation provided $45 million to assist the 24,000 existing federal websites in putting information and services online. Over a four-year period, the fund will increase to $150 million. The new federal legislation also requires "privacy impact assessments" whenever agencies introduce new technology and public comment periods when agencies add or remove information from the government website.[51]

Sponsored by Senator Joseph Lieberman of Connecticut, the goal of the legislation is to improve interagency coordination of electronic efforts and provide electronic mechanisms for citizen and group comments during federal rule-making. Through a new website called www.regulation.gov, the public can read and comment on new proposals for federal rules by the country's 160 national agencies. This includes topics such as food and drug labeling, safety standards, and industry regulation in a variety of areas.

According to one experiment in 1997 by the Department of Transportation, public comments rose dramatically when electronic-submission mechanisms were created. In 1997, when the department required written comments, there were 3,102 public reactions; however, in 2000 after electronic comments were introduced, the number of public comments rose to 62,944.[52]

Unfortunately however, fiscal pressures in 2003 led Congress to approve only $1 million of the $45 million that had been authorized in 2002.[53] Rather than fully fund the initiative, legislators cut back sharply on the interagency funding. This meant that rather than having monies that would pay for needed upgrades, e-government planners were forced to scrounge around for funds to pay for technology improvements.

But one useful aspect of the legislation is its desire to move beyond service delivery on government websites. Since past e-government efforts have been more likely to emphasize services than democratic enhancement, this bill represents a major step forward. For the first time, on a coordinated basis federal agencies are required to develop websites that improve communications with citizens and facilitate citizen involvement with policy deliberation. If implemented broadly, this legislation offers the long-term hope that technology will be used to transform the way the political system operates.

Furthermore, a commission headed by Paul Volcker has called for sweeping changes in the way the federal government is organized. Entitled *Urgent Business for America: Revitalizing the Federal Government for the 21st Century*, its report concluded that "the organization of the federal government and the operation of public programs are not good enough: Not good enough for the American people, not good enough to meet the extraordinary challenges of the century just beginning."[54] Among its recommendations are that the federal government be reorganized into a much smaller number of "mission-related executive departments," realigning House and Senate committees in line with this reorganization, reducing the number of political positions in the executive branch, raising salaries, implementing competitive outsourcing practices that advance the public interest, and giving management more flexibility in dealing with employees.

These sentiments are in keeping with the vision of e-government. By having agencies work more closely together and by having one-stop portals that integrate agency mission and service delivery, government becomes more convenient for citizens and less expensive to fund. More flexible work arrangements would make it easier to implement electronic initiatives and to reassign personnel as work flows change. Agency mission would become more closely tied to the broad functions of government. If implemented, each of these changes offers the potential of transforming government through Internet delivery systems.

Ultimately, citizen assessments of e-government will be influenced by how effective the Internet becomes at delivering services and information. The public has a bottom-line orientation that evaluates the costs and benefits of e-government. If government performs better and costs less money, citizens will become far less cynical about the public sector and more confident about government in general. Positive performance in the electronic area may even carry over to attitudes about traditional government.

The mass media will play a crucial role in educating the public about e-government. By pinpointing its virtues and its deficiencies, reporters will help determine how the public feels about online government and the use of digital technology in the public sector. As more and more resources are invested online and governments move more of their service delivery to digital systems, reporters will devote more attention to e-government. In the process, they will come to shape how ordinary people feel about the Internet era.

Coding Instructions for Government Website Content Analysis

Website name: [Enter name]
Eligible sites include: major courts, legislature, major elected officials, major departments, major agencies serving crucial functions such as health, human services, taxation, education, corrections, economic development, administration, natural resources, transportation, elections, defense, and business regulation

City/state/country: [Enter code for specific city, state, or country]

Branch: [Code 1 for executive branch, 2 for legislative branch, 3 for judicial branch, and 4 for portal page (the homepage for each state that serves as the gateway for all the websites of a particular state)]
The legislature and judiciary often have their own portal-like pages, but you still should code the branch as 2 and 3, respectively.

Particular agency: [Enter numeric code for kind of agency]
As agency titles vary from state to state, here are the various headings we used for each agency that often used different names:

Controller. *Can also be called* Auditor *or* Comptroller
Health. *Sometimes* Public Health
Human Services. *Can also be called* Social/Family/Welfare Services
Environment. *Can also be called* Environmental Quality/Protection
Higher Education. *Can also be called* Postsecondary Education, Board of Regents
Housing. *Often goes under the name of* Housing Development/Finance Authority
Motor Vehicles. *Can be difficult to find, usually part of the* Department of Transportationt, *but can also be part of varied departments like* Revenue, Secretary of State, *or a separate entity altogether. Use the search engine if you cannot find it.*
Business license. *Some states have specific licensing departments, but not often easy to find. Check the state portal page to see if there is an online service for business/vendor registration or corporate filing and work from there.*
Hunting license. *Either a distinct department (such as* Game and Fish, Wildlife) *or a subdivision of* Natural Resources.
Elderly, or Aging/Senior Services. *Often found in the* Health *or* Human Services *departments.*
Elections: *Sometimes a distinct agency, usually part of the Secretary of State's page.*

Consumer protection. *Usually found through the Attorney General's page.*
Business regulation. *Can also be called* Professional Regulation, Commerce.

Has online publications: [0 no, 1 yes]
This category includes news releases, journals, reports, studies, laws, or con-stitutions. Often major reports are in PDF format. These would count as publications.

Offers online databases: [0 no, 1 yes]
This can vary widely from statistics, charts, data to actual databases (which are like search engines except for that they are customized to retrieve specific infor-mation rather than search the entire website). Telephone directories were not in-cluded as a database. Databases are often found in the statistics, information, or publications sections of webpages.

Has audio clips: [0 no, 1 yes]
Any sound file whatsoever, whether it be in the form of a speech, radio show, website welcome, or music, such as a state song or national anthem.

Has video clips [any video file]: [0 no, 1 yes]
Examples are televised speeches/events, department commercials, picture tours, and website welcome. Could be a video clip or example of streaming video. Powerpoint presentations are not included as video clips.

Has foreign language or language translation: [0 no, 1 yes]
Can be a webpage entirely in a non-native language (e.g., "Espanol" for English-speaking countries), a link to language-translating software like Babel-fish, or having publications available in other languages.

Has commercial advertisements: [0 no, 1 yes]
Do not count as ads links to website developer and computer software available for free download such as Adobe Acrobat Reader, Netscape Navigator, or Microsoft Internet Explorer since they are necessary for viewing pages. Links to products or services available for a fee such as commercial tax-preparation soft-ware are coded as a yes as are traditional banner or pop-up ads. Ads have to be clear commercial sponsorships of a product or service. Many links on sites ap-peared to be ads, but after clicking on them, they were only promoting a partic-ular government program or event. (Links promoting state tourism often take this form.)

Has website section requiring premium fee for entry: [0 no, 1 yes]
Fee required to access particular areas on website (such as business services, ac-cess to databases, or viewing of up-to-the-minute legislation). This is not the same as a user fee for a single service. For example, you would not code a yes for the fact that some government services require payment to complete the transac-tion. This indicator is more for website sections requiring payment to enter that

area or to access a set of premium services. Code subscription service as a yes for premium fee if there is a cost associated with the subscription.

Has restricted area requiring username and password to enter: [0 no, 1 yes]
This could be access to government contract information or procurement bidding, or access to a subscription or business-services area that is password protected. Areas that have a registration requirement for a password just for information purposes (i.e., sending free email notifications or free subscriptions to the visitor) would not be considered a restricted section since the restriction is not for a general area of the website. Also, individual services that require a password, such as income tax filing, would not be coded yes because the password does not involve a general area of the website, just that specific service. Access to state employee records that require a password would not be coded yes because they are not information where the public has a right to access.

Site meets W3C disability guidelines: [0 no, 1 yes]
To evaluate this, go to website http://www.cast.org/bobby and run the evaluation test. Type in the URL for the front page of the website you are evaluating. Choose Web Content Accessibility Guidelines and set the guidelines for Priority Level One. Click on submit to determine whether the site meets this set of guidelines. There will be a report indicating whether the site meets or does not meet the guidelines.

Site meets U.S. Section 508 disability guidelines: [0 no, 1 yes]
To evaluate this, go to website http://www.cast.org/bobby and run the evaluation test. Type in the URL for the front page of the website you are evaluating. Choose U.S. Section 508 Guidelines and click on Submit to determine whether the site meets this set of guidelines. There will be a report indicating whether the site meets or does not meet the guidelines.

Has privacy policy on site: [0 no, 1 yes]
Any mention of the privacy policy of the particular website, even if it merely says the site has a privacy policy. Sometimes, a privacy policy can be found at the bottom of the page under About Us, Privacy, or Copyright Section.

Privacy policy prohibits commercial marketing of visitor information: [0 no, 1 yes]
The privacy policy states that it does not give/sell/rent visitor information to third parties.

Site prohibits creation of "cookies" or individual profiles of visitors: [0 no, 1 yes]
Many sites clearly say whether or not they enable cookies on their websites. Other sites will say that they collect no personal information whatsoever, only non-identifying information such as the domain name (e.g., @brown.edu) and the time and date when the website was visited.

Site prohibits sharing personal information without prior consent of user: [0 no, 1 yes]
The website will only share personal information (such as giving your home address) with your consent and to specifically answer your question. Passing on information to law enforcement authorities would not be coded as a yes since that is a noncommercial reason for sharing personal information.

Site can share personal information with legal authorities or law enforcement: [0 no, 1 yes]
The website will share personal information (such as giving your personal information) with legal authorities, law enforcement, or to a court under a court order.

Has visible security policy: [0 no, 1 yes]
The security policy is its own distinct link or part of the privacy policy. Once again, any mention of the policy is adequate for coding. If the site is listed as being "secure," that would be coded as having a visible security policy as well.

Security policy uses computer software to monitor network traffic: [0 no, 1 yes]
Most security policies with this feature will distinctly say that they use computer software to monitor network traffic. Aesthetic/informational features like webcounters do not count.

Has official government services available to citizens: [0 no, 1 yes]
By service, we do not mean information provided, such as being able to read a publication, access a database online, or visit a forum, but rather an actual state service where the entire transaction can occur online such as ordering a motor license, registering to vote, applying for a business permit, filing taxes online, etc. If you have to order a service online and then mail something in to execute the service, it is not fully transactable online and therefore is not considered an online service. Services must provide features where citizens/businesses apply for a service online and receive some tangible product/benefit in return. Some examples of this are ordering publications, renewing licenses, and filing taxes. Simply giving/receiving information (such as retrieving information from a database or retrieving a job application on a website) does not count. Entering social security numbers to check tax refund status would be considered a service since one is not merely entering information, but the government is providing specialized information to the web visitor. Furthermore, many websites have 'Service' links that provide no actual online services (instead just information on different programs run by the agency), so we had to check the links specifically for that purpose. Another important note is that even if the link to an online service connects the user to a different department to complete the transaction, it still counts as a service for that site. This is often seen on the state

portal pages, as they document many of the services available on all of the different agencies' sites.

Has services requiring user fee: [0 no, 1 yes]
Fee required to execute a particular service. For example, if a driver's license costs $25 and the citizen has to pay $25 online, that would not be a user fe; it is just the normal fee for the service. If, however, the agency charges a $3 processing fee on top of the $25, that would constitute a user fee.

Number of different services: [code actual number, 0 if none]
Simply count the number of online services. A site offering both hunting and fishing licenses would be coded as two services since those serve different needs and different audiences.

Allows digital signatures on transactions: [0 no, 1 yes (if not apparent, code no)]
Certain sites that allow credit card transactions will have a feature for using digital signatures.

Allows payments via credit cards: [0 no, 1 yes [if not apparent, code no)]
The website has the capability to use credit cards to complete online transactions. It is still included even if the link to use the credit cards took us to an external site to enter the information. This often is found in conjunction with services or publications that can be ordered with a credit card.

Can email department (other than webmaster): [0 no, 1 yes]
An email address for any person or division in the departmen. Even if there is not a specific email address, if there is a specific form that can be filled out to submit comments/questions/suggestions and online, this qualifies. This feature is found on the websites of large agencies and top elected officials. (The email address of the webmaster does not count.) Often located in the Contact Us section.

Email Responsiveness Test

For each state's human services department website (may be called social/family/welfare services) and for each federal agency, send the following email to the citizen contact point listed on that website (not the webmaster): "I would like to know what hours your agency is open during the week. Thanks for your help."

Response to sample email: [0 none, 1 yes]

Response time to sample email: [0 no response; 1 within one business day; 2 within two business days; 3 within three business days 4 within four business days; 5 within five business days; 6 within six business days; 7 within seven or more business days]

Has area to post comments (other than by email): [0 no, 1 yes]
These may be user surveys, bulletin boards, chat-rooms, or guestbooks. Simply having an address to email comments and suggestions does not count.

Has option for automatic email updates, newsletters, or notifications: [0 no, 1 yes]
The website gives the user the ability to sign up and register online in order to receive agency updates in such forms as newsletters, late-breaking news, and website notifications. These updates are then sent out to people who have registered to receive information or notifications.

Allows personalization of website (to tailor page to viewer interests): [0 no, 1 yes]
Can customize website to your particular interests. Often referred to as "MyNC," for example.

Has PDA or handheld access: [0 no, 1 yes]
This would include access to the government website through a pager or mobile phone, or any kind of personal digital assistant (as opposed to computer access through the Internet).

Flesch-Kincaid Grade Level Readability

From the front page of the government website, copy the text by clicking Edit, Select All, and then Edit, Copy.

Minimize this screen and open a blank Microsoft Word document.

Click Edit, Paste to move this website text into the blank Microsoft Word document.

To set your computer to display readability statistics in Microsoft Word, click on Tools, then Spelling and Grammar, then Options, and check box for "Show Readability Statistics": then click OK.

To check the text you pasted into this blank Microsoft Word document, click on Tools, then Spelling and Grammar (or the ABC icon on the ruler). Keep clicking on Ignore All until you come to the end of the text and you see the display of readability statistics. The Flesch-Kincaid Grade Level Readability number is at the bottom of this display. Round to the closest whole number and enter this one- or two-digit number into your database.

Global E-Government Rankings by Country, 2003

Singapore	46.3	Brunei	32.8
United States	45.3	East Timor	32.6
Canada	42.4	Nepal	32.5
Australia	41.5	Thailand	32.4
Taiwan	41.3	Yugoslavia	32.3
Turkey	38.3	Tunisia	32.2
Great Britain	37.7	Poland	32.2
Malaysia	36.7	Azerbaijan	32.0
Vatican	36.5	Bahamas	32.0
Austria	36.0	Palau	32.0
Switzerland	35.9	Qatar	32.0
China	35.9	Sao Tome and Principe	32.0
New Zealand	35.5	Slovenia	32.0
Finland	35.5	Somalia	32.0
Philippines	35.5	Somaliland	32.0
Denmark	35.5	Syria	32.0
Maldives	35.2	Togo	32.0
St. Lucia	35.0	Belize	32.0
Hong Kong	34.5	Uzbekistan	32.0
Germany	34.4	Chile	32.0
Netherlands	34.3	Congo (Democratic Republic)	32.0
Iceland	34.3	Cote d'Ivoire	32.0
Japan	34.2	North Korea	32.0
Tajikistan	34.0	Sweden	31.8
Belgium	34.0	South Africa	31.8
Colombia	33.9	Saudi Arabia	31.8
Czech Republic	33.8	Djibouti	31.7
France	33.8	Ukraine	31.6
Bahrain	33.8	Bulgaria	31.4
Mexico	33.7	Spain	31.3
Portugal	33.6	Peru	31.3
Israel	33.3	Cambodia	31.0
Cyprus (Republic)	33.3	Latvia	30.9
Norway	33.2	Estonia	30.9
Italy	33.2	Greece	30.9
Croatia	33.2	Armenia	30.9
Slovakia	32.8	Georgia	30.8
Romania	32.8	Jordan	30.8

Appendix II (*continued*)

Lebanon	30.7	Mauritania	28.0
Bangladesh	30.7	Moldova	28.0
Kuwait	30.7	Myanmar	28.0
Lithuania	30.5	Niue	28.0
Micronesia	30.5	Panama	28.0
Vietnam	30.5	St. Kitts	28.0
Fiji	30.4	St. Vincent	28.0
Ethiopia	30.3	Samoa	28.0
Bosnia and Herzegovina	30.1	Senegal	28.0
India	30.1	Seychelles	28.0
Belarus	30.0	Turkmenistan	28.0
Sudan	30.0	Tuvalu	28.0
Botswana	30.0	Bolivia	28.0
Haiti	30.0	Burundi	28.0
South Korea	30.0	Algeria	28.0
Hungary	29.9	Comoros	28.0
Oman	29.8	Cyprus (Turkish Republic)	28.0
Trinidad and Tobago	29.5	Egypt	28.0
Ireland	29.4	Antigua and Barbuda	28.0
Argentina	29.4	Guatemala	28.0
Gambia	29.4	Iran	28.0
Brazil	29.4	Uganda	27.7
Andorra	29.3	Malta	27.6
Russia	29.3	Burkina Faso	27.4
Nicaragua	29.2	Arab Emirates	27.4
Pakistan	29.1	Kiribati	27.0
Nigeria	29.0	Kyrgyzstan	26.9
Barbados	29.0	Dominica	26.7
Guinea-Bissau	29.0	Paraguay	26.7
Yemen	28.9	Liechtenstein	26.5
Morocco	28.9	Mauritius	26.5
Jamaica	28.9	Cape Verde	26.4
Luxembourg	28.7	Ghana	26.3
Venezuela	28.7	Cuba	26.2
Dominican Republic	28.7	Namibia	26.2
Mongolia	28.6	Zambia	26.1
Uruguay	28.5	Niger	26.0
Kazakhstan	28.4	Guyana	26.0
Albania	28.3	Kenya	25.7
Ecuador	28.3	Mozambique	25.5
Honduras	28.2	Rwanda	25.3
El Salvador	28.1	Cameroon	25.1
Afghanistan	28.0	Swaziland	25.0
Macedonia	28.0	Grenada	25.0
Mali	28.0	Monaco	24.5

Appendix II (*continued*)

San Marino	24.2	Eritrea	24.0
Libya	24.0	Indonesia	24.0
Madagascar	24.0	Iraq	24.0
Sierra Leone	24.0	Tanzania	23.3
Sri Lanka	24.0	Malawi	22.7
Tonga	24.0	Guinea	22.7
Zimbabwe	24.0	Papua New Guinea	22.4
Benin	24.0	Lesotho	21.7
Bhutan	24.0	Liberia	20.0
Central Africa	24.0	Marshall Islands	20.0
Chad	24.0	Suriname	20.0
Congo (Republic)	24.0	Vanuatu	20.0
Cook Islands	24.0	Solomon Islands	19.2
Costa Rica	24.0	Laos	19.0
Angola	24.0	Nauru	16.0
Equatorial Guinea	24.0	Gabon	16.0

Source: Author's e-government content analysis database

E-Government Best Practices

Top City Government Websites

1. Denver (http://www.denvergov.org/)
   Denver moved up from a ranking of third to the top-ranked city in the 2003 e-government study. To earn this spot, the Denver portal page is well-organized and clearly laid out. It is easy to navigate from the portal to city departments, on-line services, elected officials, or whatever else a citizen may need. In addition to offering 23 fully executable online services, all of sites within the domain offer translation into eleven languages (including Greek, two forms of Chinese, and Dutch). Publication lists and a detailed privacy policy can be found on most pages, as can email addresses to contact departments, and a comment form to provide input about the website. Denver's pages also tend to have a lower Flesch-Kincaid reading level, making them accessible to citizens of varied levels of education.

2. Charlotte (http://www.charmeck.nc.us/Home.htm)
   The city of Charlotte ranked second in the 2003 e-government study with a score of 57.3 percent. Its plethora of fully executable services, combined with a thorough privacy and security policy, helped it earn this spot. With an e-services link at the bottom of most pages, the ability to pay taxes and submit a résumé for city jobs on-line is just a click away. Also readily available are searchable databases, including one in which entering a street address allows a user to access property and demo-graphic information about a given location. Charlotte's privacy policy promises not to rent, sell, or give away personal identification information, as well as warning users that information received online might be shared with law enforcement. Fi-nally, every page has a link to a feedback/comments site that offers users the oppor-tunity to express their opinion about the website.

3. Boston (http://www.cityofboston.gov/)
   Boston ranked third in the 2003 e-government survey, moving up two spots from last year. The uncluttered portal page is both aesthetically pleasing and user-friendly. In-formation is organized into specific sections for residents, businesses and visitors. Its 43 online services, including online parking-ticket payment, access to personalized property-tax information, and a form to report a number of public works needs, can be accessed from most sites. Citizens can contact most departments by email and can submit their comments about the sites. It is also possible to personalize the site and re-ceive email updates on a variety of subjects. Most sites have thorough privacy and se-curity policies, and there are no restrictions or premium fees to view information.

4. Louisville (http://www.loukymetro.org/)
   The city of Louisville moved up significantly, placing fourth in this year's e-government ranking, up from thirtieth last year. The website is easy to navigate and many of its

pages pass Section 508 disability requirements as well as W3C disability standards. The privacy and security statement, which is at the bottom of most pages, is clearly broken up into sections that makes it obvious that the site does not share personal information and has Secure Sockets Layer protocol to safeguard personal information. Requesting a service is also easy on the website. Using a dropdown list, a user can request over twenty services, varying from snow removal to drainage problems. City officials, including the mayor, display email addresses prominently, making it easy to contact appropriate officials. Additionally, any interested user can watch an online video of the Mayor's budget address. All of these features combine make Louisville's website strong.

5. Nashville (http://www.nashville.gov/flashpgs/flashhome.htm)
    Nashville moved up significantly from its place at 33 rd last year to this year's place as the fifth ranked e-government city. The abundance of publications and databases and the 17 online services available on most sites helped push Nashville up in the ranking. The portal page displays a wealth of information, including press releases, maps, a video about the city, and the current temperature. From the dropdown menu, citizens can easily navigate to over 50 city departments. Most sites offer a way to contact the department by email and an online form to comment on the website. All the sites within the domain have the same heading, making it easy to link to the portal, the mayor's office, the metro council, online services, or a help form from any page.

Top State Government Websites

1. Massachusetts (http://www.mass.gov/portal/index.jsp)
    Massachusetts took the top ranking spot this year with a score of 46.3 percent, using an easy, accessible format to provide services for visitors, allowing Massachusetts to stand out among the other states. Most of the sites that linked to the portal (agencies, elected officials, etc.) carry the same banner heading, which provide links to the 48 online services offered, a listing of agency sites, a search option, and a link back to the portal. From this banner heading, the user can visit any site linked to the Massachusetts portal and still have immediate access to all online services. The organization of the site is clear and concise, with different services accessible in an organized fashion.

2. Texas (http://www.state.tx.us)
    Texas, the number two ranked state, has an attractive and clutter-free portal page. The website displays its fifty online services which include such vital functions as paying taxes and renewing vehicle registrations. The privacy policy is well outlined, understandable, and comprehensive, including web traffic monitoring by software to protect data security. The drop-down menus on the left side of the portal provide dozens of direct links to such varied and often-requested topics as nutrition, public records, consumer protection, and taxes. Each link offers a brief synopsis of its specific services, with information on approximately how long it may take to access the service. The listing of all state agencies from the portal page includes not only a link to the website itself but also to the "TRAIL" page—a detailed

directory listing—and to related online publications. Many of Texas's websites include at least one online service.

3. Indiana (http://www.state.in.us/)

Indiana placed third in the 2003 survey, with a total of 42.4 percent. The portal site is direct and easy to use, with several services gracing the homepage and easy accessibility to further links. Indiana has also incorporated a banner heading for each of the sites under its portal, providing links to a listing of agencies, to a "text-only" version, a link back to the portal, a link for contacting the webmaster, and a search option. With this banner, visitors can easily check the privacy policy for each site, search the entire "Access Indiana" site for a particular subject, find a contact person for questions, etc. Indiana's site also clearly solicits the preferences and experiences of its visitors, with "IN.gov Fan" stories posted on the portal.

4. Tennessee (http://www.state.tn.us)

From the catchy banner at the top of the portal to the photos of Memphis's swinging nightlife, Tennessee's websites are characterized by attractive pages and easy-to-find services. Many of the services, including applications for unemployment benefits and submission for a state job application include online demos. The Tennessee.gov portal banner heads most department and officials' websites, providing an easy link back to the portal, to on-line services, and to contact information. Tennessee websites frequently utilized multi-media options, such as the public service announcements on the Department of Labor's website and the video and audio recording of session arguments from the State Assembly's page. Many of the links to different agencies from the portal include synopses of the services each link offers. The long list of online services Tennessee offers is a well-organized page and includes the option of getting email notification when a new service is offered. A "survey" link at the bottom of each page solicits feedback from users. State agency websites are organized in the same way as the portal, so that most information is arranged in easy-to-use drop-down menus rather than scattered around the page.

5. California (http://www.state.ca.us/state/portal/myca_homepage.jsp)

California ranks fifth in the 2003 e-government research survey, proudly displaying a clear "Online Services" link right on its homepage and offering PDA accessibility. With its personalization features, an enormous number of online services available, and a clear format for navigation, California's site offers much of what one could hope for from a state site. A banner heading with a link to the portal site lines the tops of sites within the portal, providing an easy way back to the "MyCalifornia" homepage.

Top Federal Government Websites

1. Firstgov.gov Portal (http://www.firstgov.gov)

The number one ranked federal website is firstgov.gov, the federal government's portal page. From the banner at the top of the page it is possible to translate Firstgov into one of 25 other languages. The portal also includes a welcome from President George W. Bush in video format viewable on Realplayer, an archive of press releases from the federal government, and directories for contacting government officials by email, phone, and in-person. Firstgov's privacy policy protects the personal information of users by prohibiting cookies, expressly forbidding the marketing and sharing of personal information, and employing software technology to monitor web traffic and protect security. Firstgov also includes a whopping 62 services and gives users the option to receive email notification from Firstgov whenever the website is updated. The bottom of the page also includes the option to link to Firstgov or to suggest a page that Firstgov should link to.

2. Federal Communications Commission (http://www.fcc.gov/)
    The FCC ranked second in this year's federal site e-government survey. The front page boasts distinct categories for audio/visual events, commissioners, general information, and consumer information (including links to numerous services and language options). The site performs well in a majority of the measures coded for, including privacy policy, subscription to a daily report, contact information, a link to the portal, a search tool, etc. The FCC site is comprehensive, covering a range of services and easily accessible to the visitor.

3. Social Security Administration (http://www.ssa.gov)
    The Social Security Administration's website ranks third out of all federal websites and includes useful information, such as how to protect against identity theft, as well as a host of databases and statistics, including a "monthly statistic snapshot." From a menu-bar at the top, the user may select foreign-language translation in 15 separate languages. Meeting accessibility standards, the SSA's website is both WC3 and Section 508 compliant. It offers 17 separate services, which include online application forms for retirement benefits and a Medicare card. There is also a link soliciting feedback on the website's "new look" and the option to receive the SSA's electronic newsletter. A convenient drop-down menu at the top of the screen provides answers to hundreds of frequently asked questions arranged by topic. The SSA's website also features fun information, like a "kids' page" and a list of the most popular baby names for 2002.

4. Internal Revenue Service (http://www.irs.gov)
    The Internal Revenue Service's webpage, the fourth-ranked federal website, is far more inviting than the IRS's reputation. The website offers eight different services, the most prominent of which is the "e-file" service—the ability to file taxes on-line using a major credit card. It offers a host of statistics, a convenient directory of office phone numbers accessible by clicking on your area of the map of the United States, and detailed instructions on filling out forms. The website is offered in both Spanish and English, and also includes a search engine that yields accurate and up-to-date results. The IRS icon at the top of every page provides an easy link back to the portal. The website also includes a quiz called "The Perfect Match," as a fun feature for users to test their knowledge of tax benefits.

5. Library of Congress (http://www.loc.gov/)
    The Library of Congress placed fifth, with a total of 68 percent. With services from interactive sites for children to extensive international databases, this site brings to life the slogan "More than a Library" on the front page. Visitors can shop at the "LOC" Store, use the live "Ask a Librarian" help service, or browse through the exclusive online library "exhibitions"—with features that one would expect to see at

a museum, certainly not on a website! The "Global Gateway" connects visitors to a wealth of world resources, and the site has links geared toward particular categories of visitors (researchers, lawyers, teachers, children, newcomers, the blind and persons with other disabilities, etc.). In all, this is a truly impressive site that provides the visitor with much more than one would expect.

Top Global Websites

1. Singapore (http://www.ecitizen.gov.sg/)
   Singapore's wide array of sites is highlighted by an e-service portal. Virtually all Singaporean web sites have links to this clearing-house of citizen services and information. The "eCitizen" site boasts over one hundred services (the most of all web sites evaluated in this study). Other sites contain a large number of press releases, speech texts, and databases. Its relative prevalence of privacy statements, audio and video clips, and PDA access puts Singapore's websites from out in front of those of other nations. Two other notable websites are those of the Housing and Development Board and the Singapore Police Force (SPF), which offer 38 and 17 e-services, respectively.

2. United States (http://www.firstgov.gov/)
   The United States offers the most organized portal website of any country. Information and services are easy to find and logically placed. It is also the quintessential portal because it effectively agglomerates publications, databases, and services from all governmental websites and provides accessible links to them. Whereas most portal websites provide an eclectic sampling of documents and links to the home pages of government departments and ministries, the United States portal is characterized by navigable links to well-defined information and services. Except for many of the judicial websites, privacy policies are both detailed and easy to find. Two especially notable departmental websites are those of the Securities and Exchange Commission (SEC) and the Postal Service (USPS), which offer 64 and 20 online services, respectively.

3. Canada (http://www.gc.ca/)
   The Canadian web portal is another example of a site defined by functionality and ease of use. Perusal of reports, guides, and other publications is simplified by search and organization options; one can browse by title, subject, or order of release. The Canadian portal also offers updates and customization options. All of Canada's web sites but one contain a visible privacy statement, and more than half comply with disability access standards.

4. Australia (http://www.fed.gov.au/KSP/)
   Like the portals of the United States and Canada, that of Australia proves to be well-organized and designed. The chance for the user to categorize him/herself as a student, business, or other governmental organization mimics similar options at the two aforementioned portals, and provides a neat way to focus the site. Unlike the sites, of the United States and Canada, however, the Australian portal excels aesthetically in its successful incorporation of color, without diminishing its disability access compliance. In fact, an overwhelming majority of Australian websites

comply with current online disability standards. Remarkable, too, is the fact that one hundred percent of Australian websites accessed in this study contain visible privacy statements. An impressive percentage of sites also offer the option to receive email updates on events and publications.

5. Taiwan (http://www.mof.gov.tw/)

Taiwan offers an excellent array of websites. Among the data collected on Taiwan's e-government capabilities, most impressive is that one hundred percent of the sites evaluated offer services that are fully executable online. One especially distinguished site is that of the Ministry of Finance, which offers several e-services as well as a wide assortment of publications and databases. Though no single site offers more than a few e-services, the universal ability to complete a variety of transactions online continues to set Taiwan's website apart from those of other countries. Additionally, all sites evaluated contain both publications and databases, rendering each site especially useful to the average citizen.

Notes

CHAPTER 1
SCOPE, CAUSES, AND CONSEQUENCES OF ELECTRONIC GOVERNMENT

1. Michael Margolis and David Resnick, *Politics as Usual: The Cyberspace "Revolution,"* Thousand Oaks, California: Sage Publishers, 2000.

2. Philip Agre, "The Market Logic of Information," *Knowledge, Technology, and Policy*, Volume 13, number 1, 2001, pp. 67–77, and Mordecai Lee, "E-Reporting: Strengthening Democratic Accountability," IBM Center for The Business of Government Report, February 2004.

3. Federal CIO Council, "An Inventory of Federal e-Government Initiatives," January 2001, p. 15.

4. Pew Internet & American Life Project, "eGovernment Survey," June 25–August 3, 2003, unpublished data report.

5. Jamie Stiehm, "Mayor Defeated in Battle of the Bridge," *Baltimore Sun*, November 1, 2003, p. 1B.

6. David Garson, *Public Information Technology: Policy and Management Issues*, Harrisburg, Pennsylvania: Idea Group Publishing, 2003.

7. National Performance Review, "From Red Tape to Results: Creating a Government That Works Better and Costs Less," Washington, D.C.: Government Printing Office, 1993.

8. Quoted in Council for Excellence in Government, "E-Government: The Next American Revolution," Washington, D.C., undated publication, p. 1.

9. Jeffrey Seifert and Matthew Bonham, "The Transformative Potential of E-Government in Transitional Democracies," unpublished paper available online at http://www1.worldbank.org/publicsector/egov/, 2004.

10. The first definition comes from *Oxford English Dictionary Online*, New York: Oxford University Press, 2002, while the second is given in Rudi Volti, *Society and Technological Change*, 3rd ed., New York: St. Martin's Press, 1995, p. v.

11. Bruce Bimber, "The Internet and Political Transformation: Populism, Community, and Accelerated Pluralism," *Polity*, Volume 31, number 1, 1998, pp. 133–160, and "Information and Political Engagement in America: The Search for Effects of Information Technology at the Individual Level," *Political Research Quarterly*, Volume 54, number 1, 2001, pp. 53–67.

12. Quoted in Elaine Kamarck and Joseph Nye Jr., eds., *Democracy.com? Governance in a Networked World*, Hollis, New Hampshire: Hollis Publishing Company, 1999.

13. Dennis Thompson, "James Madison on Cyberdemocracy," pp. 36–37, in Kamarck and Nye, eds., *Democracy.com?*, and Bruce Bimber, *Information and American Democracy: Technology in the Evolution of Political Power*, New York: Cambridge University Press, 2003. William Galston argues, however, that the Internet may intensify fragmentation in his essay, "The Impact of the Internet on Civic

Life," pp. 40–58, in Elaine Kamarck and Joseph Nye, Jr., eds., *Governance.com: Democracy in the Information Age*, Washington, D.C.: Brookings Institution Press, 2002.

14. Jane Fountain, "The Virtual State: Toward a Theory of Federal Bureaucracy in the 21st Century," pp. 133–156, in Kamarck and Nye, eds., *Democracy.com?*. Also see Jane Fountain, "Public Sector: Early Stage of a Deep Transformation," in Brookings Task Force on the Internet, *The Economic Payoff from the Internet Revolution*, Washington, D.C.: Brookings Institution Press, 2001.

15. Fountain, "The Virtual State: Toward a Theory of Federal Bureaucracy in the 21st Century," in Kamarck and Nye, eds., *Democracy.com?*, p. 142.

16. Charles Lindblom, "The Science of 'Muddling Through,' " *Public Administration Review*, Volume 29, Spring 1959, pp. 79–88.

17. Aaron Wildavsky, *The Politics of the Budgetary Process*, 4th ed., Boston: Little, Brown, and Co., 1984.

18. Kenneth Kraemer, William Dutton, and Alana Northrop, *The Management of Information Systems*, New York: Columbia University Press, 1981; John King and Kenneth Kraemer, *The Dynamics of Computing*, New York: Columbia University Press, 1985; Kenneth Kraemer and John King, "Computing and Public Organizations, *Public Administration Review*, Volume 46 (1986), pp. 488–96, and Kenneth Kraemer and Jason Dedrick, "Computing in Public Organizations," *Public Administration Research and Theory*, Volume 7, number 1, 1997, pp. 89–112.

19. Richard Davis, *The Web of Politics: The Internet's Impact on the American Political System*, New York: Oxford University Press, 1999, pp. 146–148.

20. Margolis and Resnick, *Politics as Usual: The Cyberspace "Revolution,"* p. vii.

21. Andrew Chadwick with Christopher May, "Interaction Between States and Citizens in the Age of the Internet: E-Government in the United States, Britain and the European Union," *Governance: An International Journal of Policy, Administration, and Institutions*, Volume 16, number 2, 2003, pp. 271–300.

22. James Brian Quinn, *Intelligent Enterprise*, New York: Free Press, 1992.

23. Jane Fountain, *Building the Virtual State: Information Technology and Institutional Change*, Washington, D.C.: Brookings Institution Press, 2001.

24. Volti, *Society and Technological Change*.

25. "The First Modern Cars," website www.cybersteering.com, April 11, 2002.

26. Philip Howard and Steve Jones, eds., *Society Online: The Internet in Context*, London: Sage Publishers, 2004.

27. Amy Keller, "Most Hill Web Sites Miss Mark, Study Shows," *Roll Call*, January 28, 2002. The full study is described in Congress Online, "Assessing and Improving Capitol Hill Web Sites," 2002.

28. James Snider, "E-Democracy as Deterrence," paper delivered at the annual meeting of the American Political Science Association, San Francisco, California, August 30–September 2, 2001. Also see Andrew Chadwick with Christopher May, "Interaction Between States and Citizens in the Age of the Internet," paper presented at the annual meeting of the American Political Science Association, San Francisco, California, August 30–September 2, 2001.

29. Anthony Wilhelm, *Democracy in the Digital Age: Challenges to Political Life in Cyberspace*, New York: Routledge, 2000.

30. Thomas Beierle, "Democracy On-Line: An Evaluation of the National Dialogue on Public Involvement in EPA Decisions," Resources for the Future report, January 2002.

31. Volti, *Society and Technological Change*, New York: St. Martin's Press, 1988, p. 224.

32. Helen Margetts, *Information Technology in Central Government*, London: Routledge, 1999.

33. Barry Bozeman, *All Organizations are Public: Bridging the Gap Between Public and Private Organizational Theories*, San Francisco: Jossey-Bass, 1987.

34. William Dutton, *Society on the Line: Information Politics in the Digital Age*, New York: Oxford University Press, 1999.

35. Jane Fountain, *Information, Institutions and Governance*, Cambridge, Mass.: National Center for Digital Government, 2003, pp. 45–46.

36. Volti, *Society and Technological Change*.

37. Philip Howard, Lee Rainie, and Steve Jones, "Days and Nights on the Internet: The Impact of a Diffusing Technology," *American Behavioral Scientist*, Volume 45, November 2001, pp. 382–404.

38. Paul DiMaggio, Eszter Hargittai, Russell Neuman, and John Robinson, "Social Implications of the Internet," *Annual Review of Sociology*, Volume 27, 2001, pp. 307–336, and James Katz and Ronald Rice, *Social Consequences of Internet Use*, Cambridge, Mass.: MIT Press, 2002.

39. Jonathan Krim, "The Internet Gets Serious," *Washington Post*, June 19, 2002, p. H1.

40. Stephen Baker, "The Taming of the Internet," *Business Week*, December 15, 2003, pp. 78–82.

41. Katie Hafner, "A Change of Habits to Elude Spam's Pall," *New York Times*, October 23, 2003, p. E1.

42. Alfred Tat-Kei Ho, "Reinventing Local Governments and the E-Government Initiative," *Public Administration Review*, Volume 62, 2002, pp. 434–44, and Clay Wescott, "E-Government to Combat Corruption in the Asia Pacific Region," paper presented at the International Anti-Corruption Conference, Seoul, South Korea, May 25–28, 2003.

43. The list of features analyzed included: office phone number, office address, online publications, online database, external links to other sites, audio clips, video clips, foreign language or language translation, privacy policy, advertisements, security features, toll-free phone number, technical assistance, subject index, frequently asked questions (FAQs), disability access, services, digital signatures, credit card payments, email address, search capability, comment form, chat-room, broadcast of events, automatic email updates, push technologies that automatically send information to recipients, and personalization features.

44. Council for Excellence in Government, "E-Government," 2002. Unpublished survey report available online at www.excelgov.org.

45. Council for Excellence in Government, "The New E-Government Equation," April 2003. Unpublished survey report available online at www.excelgov.org.

46. More detailed descriptions of this content analysis of e-government websites at various levels of the public sector are available online at www.InsidePolitics.org.

CHAPTER 2
BUREAUCRATIC, FISCAL, AND POLITICAL CONTEXTS

1. Charles Perrow, *Complex Organizations: A Critical Essay*, 3rd ed., New York: Random House, 1986, and James Q. Wilson, *Bureaucracy: What Government Agencies Do and Why They Do It*, New York: Basic Books, 1989.

2. John Huber, Charles Shipan, and Madelaine Pfahler, "Legislatures and Statutory Control of Bureaucracy," *American Journal of Political Science*, Volume 45, number 2, April 2001, pp. 330–45.

3. Michael Alvarez and John Brehm, "Speaking in Two Voices: American Equivocation about the Internal Revenue Service," *American Journal of Political Science*, Volume 42, number 2, April 1998, pp. 418–19.

4. Paul Abramson, *Political Attitudes in America*, San Francisco: W. H. Freeman, 1983.

5. Daniel Hallin, *The "Uncensored War,"* New York: Oxford University Press, 1986, and Gladys and Kurt Lang, *The Battle for Public Opinion*, New York: Columbia University Press, 1983.

6. Michael Isikoff, *Uncovering Clinton*, New York: Crown, 1999.

7. William Jacoby, "Issue Framing and Public Opinion on Government Spending," *American Journal of Political Science*, Volume 44, number 4, October 2000, pp. 750–67.

8. David Osborne and Peter Plastrick, *Banishing Bureaucracy: The Five Strategies for Reinventing Government*, New York: Addison-Wesley, 1992, p. 17. Also see David Osborne and Ted Gaebler, *Reinventing Government: How the Entrepreneurial Spirit is Transforming the Public Sector*, New York: Dutton/Plume Publishers, 1992.

9. Quote from President Bill Clinton's announcement of National Performance Review, March 3, 1993.

10. Kuno Schedler and Maria Christina Scharf, "Exploring the Interrelations Between Electronic Government and the New Public Management," unpublished paper, 2003.

11. Douglas Holmes, *EGov: E-Business Strategies for Government*, London: Nicholas Brealey Publishing, 2001.

12. Abraham McLaughlin, "The Bush Team Brings Very Corporate Values," *Christian Science Monitor*, January 9, 2001, p. 1.

13. Ezra Suleiman, *Dismantling Democratic States*, Princeton: Princeton University Press, 2003; see especially chapter 3.

14. Joel Aberbach, Robert Putnam, and Bert Rockman, *Bureaucrats and Politicians in Western Democracies*, Cambridge, Mass.: Harvard University Press, 1981.

15. For a similar argument using 2001 data, see Jae Moon and Eric Welch, "Same Bed, Different Dreams?" A Comparative Analysis of Citizen and Bureaucrat Perspectives on E-Government," unpublished paper, 2003.

16. National Performance Review, "From Red Tape to Results: Creating a Government That Works Better and Costs Less," Washington, D.C.: Government Printing Office, 1993.

17. Jane Fountain, *Information, Institutions and Governance*, Cambridge, Mass.: National Center for Digital Government, 2003, pp. 45–46.

18. Maria Christina Scharf, "Knowledge Flows and the Use of Internet-Related In-

formation Technologies in Public Sector Organizations: A Comparative Case Study," paper presented at the annual meeting of the American Political Science Association, Philadelphia, Penn., August 27–31, 2003.

19. Jack Walker, "The Diffusion of Innovations Among the American States," *American Political Science Review*, Volume 63, number 3, September 1969, pp. 880–99.

20. Michael Lipsky, *Street-Level Bureaucracy: Dilemmas of the Individual in Public Services*, New York: Russell Sage Foundation, 1980, and Stuart Bretschneider, "Red Tape, Administrative Process, and Extreme Behavior: A Top Down Approach to Measuring the Impact of Red Tape," paper presented at the second national Conference on Public Management, University of Wisconsin, Madison, 1993.

21. Charles Hinnant, "Information Technology and Organizational Control: Examining the Management of IT in an Era of E-Government," paper presented at the annual meeting of the American Political Science Association, Boston, Mass., August 29–September 1, 2002.

22. John Kamensky, "The Best-Kept Secret in Government: How the NPR Translated Theory into Practice," pp. 58–96, in Patricia Ingraham, James Thompson, and Ronald Sanders, eds., *Transforming Government: Lessons from the Reinvention Laboratories*," San Francisco: Jossey-Bass, 1998. Also see Hal Rainey, "Ingredients for Success: Five Factors Necessary for Transforming Government," pp. 147–72, in Ingraham, Thompson, and Sanders, eds., *Transforming Government*, and Patrick Dunleavy, Helen Margetts, Simon Bastow, Jane Tinkler, and Hala Yared, "Policy Learning and Public Sector Information Technology," paper prepared for delivery at the annual meeting of the American Political Science Association, San Francisco, Cal., August 30–September 2, 2001.

23. Louis Tornatzky and K. Klein, "Innovation Characteristics and Innovation Adoption-Implementation: A Meta-Analysis of Findings," *IEEE Transactions on Engineering Mangement*, Volume 29, number 1, 1982, pp. 28–45. Also see Louis Tornatzky and Mitchell Fleischer, *The Process of Technological Innovation*, Lexington, Mass.: Lexington Books, 1990.

24. Daniel Bugler and Stuart Bretschneider, "Technology Push or Program Pull," pp. 275–93, in Barry Bozeman, ed., *Public Management: The State of the Art*, San Francisco: Jossey-Bass, 1993, and Charles Hinnant, "Perceptions of Technology Acceptance: Examining the Adoption of E-Government Technologies by State Agencies," paper presented at the annual meeting of the American Political Science Association, San Francisco, Cal., August 30–September 2, 2001.

25. Jae Moon, "The Evolution of E-government among Municipalities: Rhetoric or Reality?" *Public Administration Review*, Volume 62, number 4, 2002, pp. 424–33. For a related argument, also see Jae Moon and Stuart Bretschneider, "Can State Government Actions Affect Innovation and Its Diffusion?", *Technological Forecasting and Social Change*, Volume 54, 1997, pp. 57–77.

26. Charles Hinnant, "Perceptions of Technology Acceptance: Examining the Adoption of E-Government Technologies by State Agencies," paper presented at the annual meeting of the American Political Science Association, San Francisco, Cal., August 30–September 2, 2001, and Charles Kaylor, Randy Deshazo, and David Van Eck, "Gauging E-Government: A Report on Implementing Services Among American Cities," *Government Information Quarterly*, Volume 18, 2001, pp. 293–307.

27. Diana Gant, Jon Gant, and Craig Johnson, "State Web Portals: Delivering and

Financing E-Service," Washington, D.C.: The PricewatehouseCoopers Endowment for the Business of Government, January 2002.

28. Lori Sharn, "Driving E-Government," *Catalyst*, 2001, p. 34.

29. Ellen McCarthy, "Government Sites Draw Web Traffic," *Washington Post*, January 9, 2002.

30. Bryan Jones, Frank Baumgartner, and James True, "Policy Punctuations: U.S. Budget Authority, 1947–1995," *Journal of Politics*, Volume 60, number 1, February 1998, pp. 1–33.

31. Aaron Wildavsky, *The Politics of the Budgetary Process*, 4th ed., Boston: Little, Brown, and Co., 1984.

32. See Gant, Gant, and Johnson, "State Web Portals," and Susan MacManus, "Taxing and Spending Politics: A Generational Perspective," *Journal of Politics*, Volume 57, number 3, August 1995, pp. 607–29.

33. Information drawn from NIC Corporate Web Site at www.nicusa.com, 2002.

34. Alan Albarran and David Goff, eds., *Understanding the Web: The Social, Political, and Economic Dimensions of the Internet*, Ames: Iowa University Press, 2000.

35. Rebecca Raney, "An Effort to Untangle the Government's Haphazard Approach to the Acquisiton of Computer Systems," *New York Times*, July 8, 2002, p. C4.

36. Ibid.

37. Gant, Gant, and Johnson, "State Web Portals."

38. Peter Hart and Bob Teeter, Survey of 408 Government Workers, conducted February 20–28, 2003. Unpublished study; results provided to the author.

39. Dov Zakheim and Jeffrey Ranney, "Matching Defense Strategies to Resources: Challenges for the Clinton Administration," *International Security*, Volume 18, number 1, summer 1993, pp. 51–78.

40. Patrick Dunleavy, Helen Margetts, Simon Bastow, Jane Tinkler, and Hala Yared, "Policy Learning and Public Sector Information Technology," paper delivered at the annual meeting of the American Political Science Association, San Francisco, Cal., August 30–September 2, 2001.

41. Jerry Webman, "UDAG: Targeting Urban Economic Development," *Political Science Quarterly*, Volume 96, number 2, summer 1981, pp. 189–207, and Peter Eisinger, "Do the American States Do Industrial Policy?", *British Journal of Political Science*, Volume 20, number 4, October 1990, pp. 509–35.

42. Spencer Hsu, "New Site Streamlines Online Government," *Washington Post*, September 23, 2000, p. A5.

43. Ibid.

44. Ellen Nakashima, "FirstGov Web Site Again Is Clicking," *Washington Post*, March 1, 2002, p. A23.

45. Ramona McNeal, Caroline Tolbert, Karen Mossberger, and Lisa Dotterweich, "Innovating in Digital Government in the American States," *Social Science Quarterly*, Volume 84, number 1, March 2003, pp. 52–70.

46. Paul Ferber, Franz Foltz, and Rudy Pugliese, "The Politics of State Legislature Websites," paper presented at the annual meeting of the American Political Science Association, Boston, Mass., August 29–September 1, 2002.

47. Rebecca Raney, "An Effort to Untangle the Government's Haphazard Approach to the Acquisiton of Computer Systems," *New York Times*, July 8, 2002, p. C4.

48. James Jefferson, "Arkansas E-Government Services Rank High in Surveys," Associated Press State and Local Wire, September 19, 2002, and Todd Richmond, "Doyle Attacks E-Government," Associated Press State and Local Wire, September 28, 2002.

49. Richmond, "Doyle Attacks E-Government."

50. Lydia Polgreen, "Albany Web Site Puts City in the E-Basement," *Albany Times Union*, October 31, 2001, p. B1.

51. Phillip Henderson, "Technocratic Leadership," in Phillip Henderson, ed., *The Presidency Then and Now*, Lanham, Md.: Rowman & Littlefield Publishers, 2000, pp. 219–39.

52. John Kingdon, *Agendas, Alternatives, and Public Policies*, Boston: Little, Brown, 1984.

53. Jefferson, "Arkansas E-Government Services Rank High in Surveys."

54. Ibid.

55. Richmond, "Doyle Attacks E-Government."

56. Ibid.

57. Adam Clymer, "U.S. Revises Sex Information, and a Fight Goes On," *New York Times*, December 27, 2002, p. A15.

58. Ibid.

59. Tom Diemer, "Security Concerns Keep Data Off Web," *Cleveland Plain Dealer*, November 5, 2001, p. A1, and Elisabeth Bomiller, "A Nation Challenged: Flow of Information," *New York Times*, October 7, 2001, p. 1B.

CHAPTER 3
THE CONTENT OF AMERICAN GOVERNMENT WEBSITES

1. Irina Ceaparu and Ben Shneiderman, "Improving Web-based Civic Information Access: A Case Study of the 50 US States," report available online at http://www.cs.umd.edu/hcil/pubs/presentations/4.

2. Taylor Nelson Sofres, "Government Online: An International Perspective," November 2002. Unpublished paper.

3. Center for Digital Government, *2002 Digital State Survey*, 2003, report available online at www.centerdigitalgov.com.

4. Steve Towns, "Digital Dogfight," *Government Technology*, November, 2002, report available online at www.govtech.net.

5. Accenture, "eGovernment—More Customer Focused than Ever Before," 2002, report available online at www.accenture.com.

6. Steve Rohleder, "Global eGovernment: Charting the Path to Progress," April 8, 2003, speech available online at www.accenture.com.

7. Chris Demchak, Christian Friis, and Todd LaPorte, "Webbing Governance: National Differences in Constructing the Face of Public Organizations," pp. 45–77, in G. David Garson, ed., *Handbook of Public Information Systems*, New York: Marcel Dekker Publishers, 2000.

8. United Nations Division for Public Economics and Public Administration and American Society for Public Administration, "Benchmarking E-government: A Global Perspective," May, 2002, report available online at www.unpan.org. Also see

United Nations, *World Public Sector Report 2003: E-Government at the Crossroads*, New York: United Nations, 2003.

9. Paul Ferber, Franz Foltz, and Rudy Pugliese, "The Politics of State Legislature Websites," paper presented at the annual meeting of the American Political Science Association, Boston, Mass., August 29–September 1, 2002; Jody and Bryan Fagan, "Citizens' Access to On-Line State Legislative Documents," *Government Information Quarterly*, Volume 18, 2001, pp. 105–121; Matt Carter and Ryan Turner, "New Study Assesses State Legislative Online 'Entry Points,' " March 19, 2001, unpublished report.

10. Congress Online Project, "Congress Online 2003: Turning the Corner on the Information Age," September 2003, unpublished report available online at www .congressonlineproject.org.

11. Juliet Musso, Christopher Weare, and Matt Hale, "Designing Web Technologies for Local Governance Reform: Good Management or Good Democracy?," *Political Communication*, Volume 17, January–March, 2000, pp. 1–19.

12. Joint Legislative Audit Committee, "Local E-Government Services," December, 2001, unpublished report.

13. For an effort to think systematically about the criteria for website evaluation, see Charles McClure, Timothy Sprehe, and Kristen Eschenfelder, "Performance Measures for Federal Agency Websites," October 1, 2000, unpublished report.

14. More detailed discussion of these sites can be found in e-government reports online at www.InsidePolitics.org.

15. M. Jae Moon, "The Evolution of E-Government Among Municipalities: Rhetoric or Reality?" *Public Administration Review*, Volume 62, number 4, 2002, pp. 424–433.

16. Similar studies were undertaken on global e-government, where we investigated e-government in the 198 nations around the world. Summaries of those studies are reported in chapter 9 and are available online at www.InsidePolitics.org. See Darrell M. West, "Global E-Government," unpublished 2001, 2002, and 2003 reports.

17. Anna Brannen, "E-Government in California: Providing Services to Citizens Through the Internet," *Spectrum: The Council of State Governments*, spring 2001, pp. 6–10.

18. National Performance Review, "From Red Tape to Results: Creating a Government That Works Better and Costs Less," Washington, D.C.: Government Printing Office, 1993.

19. Irwin Kirsch, Ann Jungeblut, Lynn Jenkins, and Andrew Kolstad, *Adult Literacy in America*, Washington, D.C.: National Center for Education Statistics, U.S. Department of Education, September 1993.

20. Louis Tornatzky and K. Klein, "Innovation Characteristis and Innovation Adoption-Implementation: A Meta-Analysis of Findings," *IEEE Transactions on Engineering Management*, Volume 29, number 1, 1982, pp. 28–45.

21. Ben Shneiderman, "Universal Usability," *Communications of the ACM*, May 2000, Volume 43, number 5, pp. 85–91.

22. Martin Gould, Switzer Seminar Series Remarks, Michigan State University, October 4, 2001.

23. Jim Ellison, "Assessing the Accessibility of Fifty Federal Government Web Pages," unpublished paper, 2004.

24. Sarah Horton, "Eye-Popping Graphics Can Spice Up Web Sites, But They Also Create Barriers," *New York Times*, June 10, 2002, p. C4.

25. John Miller, "English Is Broken Here," *Policy Review*, number 79, September–October 1996.

26. Also see Genie Stowers, "Becoming Cyberactive: State and Local Governments on the World Wide Web," *Government Information*, Volume 16, number 2, 1999, pp. 111–127.

27. Daniel Bugler and Stuart Bretschneider, "'Technology Push or Program Pull,'" pp. 275–293, in Barry Bozeman, ed., *Public Management: The State of the Art*, San Francisco: Jossey-Bass, 1993, and Charles Hinnant, "Perceptions of Technology Acceptance: Examining the Adoption of E-Government Technologies by State Agencies," paper presented at the annual meeting of the American Political Science Association, San Francisco, Cal., August 30–September 2, 2001.

28. S.v. "transformation" in the *Oxford English Dictionary Online*, New York: Oxford University Press, 2002.

CHAPTER 4
EXPLAINING E-GOVERNMENT PERFORMANCE

1. See Paul Baker and Costas Panagopoulos, "Great Expectations: The Promise of Digital Government in the American States," paper delivered at the annual meeting of the American Political Science Association, Philadelphia, Penn., August 28–31, 2003.

2. The Council of State Governments, *The Book of the States*, 2003 Edition, Volume 35, Lexington, Ky., 2003, pp. 308–310.

3. Peverill Squire, "Legislative Professionalization and Membership Diversity in State Legislatures," *Legislative Studies Quarterly*, Volume 17, number 1, 1992, pp. 69–79.

4. Ramona McNeal, Caroline Tolbert, Karen Mossberger, and Lisa Dotterweich, "Innovating in Digital Government in the American States," *Social Science Quarterly*, Volume 84, number 1, March 2003, pp. 52–70.

5. The Council of State Governments, *The Book of the States*, 2003 Edition, Volume 35, Lexington, Ky., 2003, p. 368.

6. Ibid., pp. 113–114.

7. Virginia Gray, "Innovation in the States," *American Political Science Review*, Volume 67, 1973, pp. 1174–1185.

8. The data on Internet usage by percentage of state population were taken from the U.S. Department of Commerce, National Telecommunications and Information Administration, "A Nation Online: How Americans Are Expanding Their Use of the Internet," February 2002, pp. 8–9. This report is available online at www.ntia.doc.gov/ntiahome/. The data on the percentage of government websites within each state where the public can communicate or transact business with the governing using Internet, e-mail, or other computer-based systems are derived from a 2002 U.S. Census report on government participation in e-government activities. These data are available online at www.Census.gov/prod/2003pubs/.

9. Council of State Governments, *Book of the States*, p. 440.

10. Ibid., pp. 384–85.

11. Theodore Lowi, "Four Systems of Policy, Politics and Choice," *Public Administration Review*, Volume 33, 1972, pp. 298–310.

12. McNeal et al., "Innovating in Digital Government," pp. 52–70.

CHAPTER 5
THE CASE OF ONLINE TAX FILING

1. Arlene Weintraub, "Coming Soon: Talking Tax Forms," *Business Week*, September 16, 2002, p. 60.

2. *New York Times*, "Electronic Tax Filing," January 31, 2002, p. A21.

3. Darrell M. West, "Online Tax Filing," unpublished report, June 2002.

4. *USA Today*, "Bush Proposed Free E-Filing, But Tax-Preparation Firms Hate That Idea," April 15, 2002, p. 8A.

5. Neil Downing, "Electronic Tax Returns Free and Easier, IRS Says," *Providence Journal*, October 21, 2003, p. E1.

6. Jeff Gelles, "Have Your Taxes Prepared for Free, But Watch for Extras," *Philadelphia Inquirer*, January 26, 2003, p. E1.

7. Neil Downing, "You'll Still Pay to File R.I. Return Electronically," *Providence Journal*, November 3, 2002, p. F3.

8. Gelles, "Have Your Taxes Prepared for Free," p. E1.

9. Downing, "Electronic Tax Returns Free and Easier," p. E1.

10. *USA Today*, "Bush Proposed Free E-Filing," p. 8A.

11. The 2001 numbers come from United States General Accounting Office, "Assessment of IRS 2001 Tax Filing System," GAO-02-144, December 2001, p. 14. The 2002 and 2003 numbers are reported in an Internal Revenue Service press release, "Strong Filing Season Produces E-file Records," May 2004.

12. Ibid., p. 31.

13. Steven Cohen and William Eimicke, "The Use of the Internet in Government Service Delivery," The PriceWaterhouseCoopers Endowment for the Business of Government E-Government Series, February 2001.

14. United States General Accounting Office, "Assessment of IRS 2001 Tax Filing System," pp. 15–19.

15. Cohen and Eimicke, "Use of the Internet in Government Service Delivery."

16. United States General Accounting Office, "Assessment of IRS 2001 Tax Filing System," pp. 30–32.

17. West, "Online Tax Filing."

18. Ibid.

19. Ibid.

20. Ibid.

21. *Federal Computer Week*, April 2004, unpublished survey.

22. West, "Online Tax Filing."

23. Jonathan Krim, "The Internet Gets Serious," *Washington Post*, June 19, 2002, p. H1.

24. Jeffrey Abramson, Christopher Arterton, and Gary Orren, *The Electronic Com-*

monwealth: The Impact of New Media Technologies on Democratic Politics, New York: Basic Books, 1988.

Chapter 6
Public Outreach and Responsiveness

1. National Performance Review, "From Red Tape to Results: Creating a Government That Works Better and Costs Less," Washington, D.C.: Governmental Printing Office, 1993.

2. Robert Dahl, *On Democracy*, New Haven: Yale University Press, 1998, and idem, *Who Governs: Democracy and Power in an American City*, New Haven: Yale University Press, 1961.

3. Marion Just, Ann Crigler, Timothy Cook, Dean Alger, Montague Kern, and Darrell M. West, *Crosstalk*, Chicago: University of Chicago Press, 1996.

4. Carole Pateman, *Participation and Democratic Theory*, New York: Cambridge University Press, 1970.

5. Jeffrey Abramson, Christopher Arterton, and Gary Orren, *The Electronic Commonwealth: The Impact of New Media Technologies on Democratic Politics*, New York: Basic Books, 1988.

6. Michael Schrage, " 'ET' Phones Hill—With a TV Poll Of Ohio Viewers' Instant Views," *Washington Post*, March 10, 1983, p. E1.

7. Tony Schwartz, "TV Notebook: Armchair Quarterbacks Get to Coach Real Game," *New York Times*, June 27, 1980, p. C31.

8. C. Gerald Fraser, "Audience's Response to Direct the Course of a TV Show," *New York Times*, November 15, 1980, p. 48.

9. Ken Freed, "When Cable Went Qubist," *Media Visions Journal*, 2000. Online journal at www.media-visions.com. Also see Ronald Hicks and Michael Dunne, "Do-It-Yourself Polling: A Case Study and a Critique," *Newspaper Research Journal* Volume 1, 1980, 46–52, and Chuck Hakes, "Test of an Automatic Telephone Interviewing Device vs. Live Television," ibid., 13–17.

10. Abramson, Arterton, and Orren, *Electronic Commonwealth*.

11. Don Oldenburg, "Spam and Ughs: As Unsavory E-Mail Bloats the In-Box, Fed-Up Recipients Turn to the Law," *Washington Post*, September 2, 2002, p. C1.

12. Jonathan Krim, "The Internet Gets Serious," *Washington Post*, June 19, 2002, p. H1.

13. Joe Light, "Lawmakers' Websites Receive Poor Grades," *The Hill*, June 26, 2002, and Michael Gerber, "Homeland Security Association Opening New Lobbying Frontier," *The Hill*, October 23, 2002, p. 29.

14. Thomas Parackal, "Web Site Personalization," undated document found at www.stylusinc.com, July 2, 2002.

15. For a study of search features of state legislative websites, see Jody and Bryan Fagan, "Citizens' Access to On-line State Legislative Documents," *Government Information Quarterly*, Volume 18, 2001, pp. 105–121.

16. The 2003 e-government content analysis did not tabulate the percentage of sites that were searchable or that offered broadcast features.

17. Cass Sunstein, *Republic.com*, Princeton: Princeton University Press, 2001.

18. *Business Week*, "Anybody Out There?" August 4, 2003, p. 14.

CHAPTER 7
CITIZEN USE OF E-GOVERNMENT

1. Darrell M. West, *The Rise and Fall of the Media Establishment*, Boston: Bedford/St. Martin's Press, 2001, pp. 28, 131.

2. Tina Rowan Mekeal, "Some Early Day Finney County History and Folks," www.eng.uci.edu/students/mpontius/hartley/16-20_ea.html, undated.

3. Ibid.

4. Reported on www.ideafinder.com/history/inventions/. See entry for Invention Facts and Telephone.

5. Pippa Norris, *Digital Divide: Civic Engagement, Information Poverty, and the Internet Worldwide*, New York: Cambridge University Press, 2001.

6. National Research Council, *A Study of the Effect of Automatic Sequence Clothes Washing Machines on Individual Sewage Disposal Systems*, Washington, D.C.: National Research Council, 1956.

7. Pablo Iannone, ed., *Contemporary Moral Controversies in Technology*, New York: Oxford University Press, 1987, and Ian Barbour, *Ethics in an Age of Technology*, San Francisco: Harper San Francisco, 1993.

8. J. Robert Oppenheimer, *The Constitution of Matter*, Eugene, Ore.: State System of Higher Education, 1956.

9. Lawrence Badash, *Scientists and the Development of Nuclear Weapons: From Fission to the Limited Test Ban Treaty, 1939–1963*, Atlantic Highlands, N.J.: Humanities Press, 1995.

10. Rudi Volti, *Society and Technological Change*, New York: St. Martin's Press, 1988.

11. Mary Wollstonecraft Shelley, *Frankenstein, or the Modern Prometheus*, New York: Illustrated Editions Company, 1932.

12. Ian Wilmut, Keith Campbell, and Colin Tudge, *The Second Creation: Dolly and the Age of Biological Control*, New York: Farrar, Straus and Giroux, 2000.

13. Victoria Griffith, "U.S. Plan for Total Cloning Ban," *Financial Times*, November 13, 2002, p. 12.

14. Walter Urbain, *Food Irradiation*, Orlando, Fla.: Academic Press, 1986.

15. Food Irradiation Campaign Press Release, "Scientists in Row Over Safety of Irradiated Foods," June 25, 2002, http://ourworld.compuserve.com/homepages/foodcomm/IRAD_prs.HTM.

16. Charles Daniel, *Lords of the Harvest: Biotech, Big Money, and the Future of Food*, Cambridge, Massachusetts: Perseus Publishers, 2001.

17. Pew Research Center, "Public Makes Distinctions on Genetic Research," released April 9, 2002. National survey conducted with 2,002 American adults between February 25 and March 10, 2002.

18. Food Irradiation Campaign, "Attitudes to Food Irradiation in Europe," September 2002, p. 2.

19. Pew Research Center, "Millennium Survey," released July 3, 1999. National survey conducted with 789 American adults between April 6 and May 6, 1999.

20. Ibid.

21. Lemelson-MIT Invention Index Study, November, 2003, p. 23, unpublished report made available by authors.

22. Council for Excellence in Government, "E-Government: The Next American Revolution," September 2000, unpublished survey report available online at www.excelgov.org.

23. Ibid.

24. Council for Excellence in Government, "The New E-Government Equation," April 2003, unpublished survey report available online at www.excelgov.org.

25. Darrell M. West, "State and Federal E-Government in the United States," September 2002, unpublished report available online at www.InsidePolitics.org.

26. Council for Excellence in Government, "The New E-Government Equation," p. 20.

27. Chris Seper, "Glenn Center on Guard for Hackers," *Cleveland Plain Dealer*, February 21, 2002, p. C1.

28. Jon Swartz, "Experts Fear Cyberspace Could Be Terrorists' Next Target," *USA Today*, October 9, 2001, p. 1B.

29. Susan Stellin, "Reports of Hackers Are on the Rise," *New York Times*, January 21, 2002, p. C6.

30. Paul Abramson, *Political Attitudes in America: Formation and Change*, San Francisco: W. H. Freeman, 1983.

31. Unlike the pattern at the state and national levels, local e-government users tend to be members of a minority. Twenty-two percent of those who have visited local sites are part of a minority, compared to 15 percent of state government website users.

32. The 2001 national survey interviewed 961 adults across the country, including 155 individuals who reported they were Internet users. The survey was conducted November 12–19, 2001. It had a margin of error of about plus or minus 3 percent. The survey was conducted by Peter Hart and Robert Teeter of Washington, D.C.

33. Council for Excellence in Government, "The New E-Government Equation," p. 12.

34. Sidney Verba and Norman Nie, *Participation in America*, New York: Harper & Row, 1972.

35. For a similar result, see Bruce Bimber, "Information and Political Engagement in America: The Search for Effects of Information Technology at the Individual Level," *Political Research Quarterly*, Volume 54, number 1, March 2001, pp. 53–67.

36. Robert Reich, *The Work of Nations: Preparing Ourselves for 21st Century Capitalism*, New York: Vintage, 1992.

CHAPTER 8
TRUST AND CONFIDENCE IN E-GOVERNMENT

1. National Performance Review, "From Red Tape to Results: Creating a Government that Works Better and Costs Less," Washington, D.C.: Government Printing Office, 1993.

2. Pippa Norris, *Digital Divide? Civic Engagement, Information Poverty and the Internet Worldwide*, New York: Cambridge University Press, 2001.

3. The Gallup Organization, "Effects of Year's Scandals Evident in Honesty and Ethics Ratings," December 4, 2002, www.gallup.com.

4. National Performance Review, "From Red Tape to Results."

5. C. Thomas, "Maintaining and Restoring Public Trust in Government Agencies and Their Employees," *Administration and Society*, Volume 30, 1998, pp. 166–193.

6. Warren Miller and Santa Traugott, *American National Election Studies Data Sourcebook, 1952–1986*, Cambridge, Mass.: Harvard University Press, 1989.

7. Joseph Nye, Philip Zelikow, and David King, *Why People Don't Trust Government*, Cambridge, Mass.: Harvard University Press, 1997.

8. Miller and Traugott, *American National Election Studies*.

9. The Gallup Organization, "Effects of Year's Scandals."

10. Daniel Hallin, *The Uncensored War: The Media and Vietnam*, New York: Oxford University Press, 1986.

11. Peter Braestrup, *Big Story: How the American Press and Television Reported and Interpreted the Crisis of Tet 1968 in Vietnam and Washington*, Boulder: Westview Press, 1977.

12. Gladys and Kurt Lang, *The Battle for Public Opinion: The President, the Press, and the Polls During Watergate*, New York: Columbia University Press, 1983.

13. Charles O. Jones, *The Trusteeship Pesidency: Jimmy Carter and the United States Congress*, Baton Rouge: Louisiana State University Press, 1988.

14. Charles O. Jones, ed., *The Reagan Legacy: Promise and Performance*, Chatham, N.J.: Chatham House Publishers, 1988.

15. Michael Isikoff, *Uncovering Clinton: A Reporter's Story*, New York: Crown Publishers, 1999.

16. Nye, Zelikow, and King, *Why People Don't Trust Government*.

17. These terrorist attacks boosted citizen confidence in government over the short run; however, the further away those events became, the less trusting citizens became of government performance.

18. Pew Research Center, "News Media's Improved Image Proves Short-Lived," August 4, 2002, http://people-press.org.

19. Jeffrey Seifert and Eric Peterson, "The Promise of all Things E?" paper presented at the annual meeting of the American Political Science Association, San Francisco, California, August 30–September 2, 2001.

20. National Performance Review, "From Red Tape to Results."

21. Similar to other surveys, citizens in this national sample were cynical about and disengaged from the political process. Only 30 percent said they trusted the government in Washington to do what is right most of the time, while 69 percent felt you could trust it only some of the time or never. Twenty-six percent reported they have quite a lot of confidence in the federal government, compared to 30 percent who felt that way about state government and 31 percent who believed it of local government. Fifty-four percent thought the government today is effective at solving problems and helping people. Twenty-seven percent said they are fairly active in politics and government, while 32 say they are somewhat active and 41 percent indicate they are not too active in politics and government.

22. Paul Abramson, *Political Attitudes in America: Formation and Change*, San Francisco: W. H. Freeman, 1983.

23. Sidney Verba and Norman Nie, *Participation in America*, New York: Harper & Row, 1972.

24. The same was true for state web usage and local web usage. Using a regression analysis and controlling for the same factors, there were no significant relationships between either one of these items and trust in government, confidence in government, or belief that government is effective.

25. There was no significant relationship between changes in beliefs about government effectiveness and state or local website usage, controlling for the same factors as in the federal e-government model.

26. Phillip Henderson, "Technocratic Leadership," in Phillip Henderson, ed., *The Presidency Then and Now*, Lanham, Md.: Rowman & Littlefield Publishers, 2000, pp. 219–239.

27. John Kingdon, *Agendas, Alternatives, and Public Policies*, Boston: Little, Brown, 1984.

28. Suzanne Garment, *Scandal: The Crisis of Mistrust in American Politics*, New York: Times Books, 1991.

29. Erik Lords, "Web Site Tab of $1.4 Million Riles Council," *Detroit Free Press*, November 5, 2002.

30. Robert Salladay, "Davis Ousts Aide in Oracle Flap," *San Francisco Chronicle*, May 3, 2002, p. A1.

31. Associated Press, "Another Davis Aide Resigns in Midst of Contracts Controversy," May 30, 2002.

32. Associated Press, "Chief Information Officer Resigns Amid Scandal," December 4, 2002.

33. Andrew Park, "Outsourcing: Look Who's Out of Sorts," *Business Week*, December 29, 2003, pp. 46–48.

34. Milton Lodge, Kathleen McGraw, and Patrick Stroh, "An Impression-Driven Model of Candidate Evaluations," *American Political Science Review*, Volume 87, number 2, June 1989, pp. 399–419.

CHAPTER 9
GLOBAL E-GOVERNMENT

1. Eric Welch and Wilson Wong, "Global Information Technology Pressure and Government Accountability: The Mediating Effect of Domestic Context on Website Openness," *Journal of Public Administration Research and Theory*, Volume 11, number 4, 2001, pp. 509–538, and Gregory Curtin, Michael Sommer, and Veronika Vis-Sommer, eds., *The World of E-Government*, Binghamton, N.Y.: Haworth Press, 2004.

2. OECD e-Government Studies, *The e-Government Imperative*, Paris, France: Organization for Economic Co-Operation and Development, 2003.

3. Funding for the 2001 global e-government research reported in this chapter was provided by World Markets Research Centre, a London consulting company.

4. Pippa Norris, *Digital Divide? Civic Engagement, Information Poverty and the Internet Worldwide*," New York: Cambridge University Press, 2001.

5. Eszter Hargittai, "Weaving the Western Web: Explaining Differences in Internet

Connectivity among OECD Countries," *Telecommunications Policy*, Volume 23, 1999, pp. 701–718.

6. Sampsa Kiiski and Matti Pohjola, "Cross-Country Diffusion of the Internet," *Information Economics and Policy*, Volume 14, 2002, pp. 297–310.

7. Pippa Norris, *Digital Divide?*

8. Results from Taylor Nelson Sofres report undertaken by Wendy Mellor, Victoria Parr, and Michelle Hood, "Government Online: An International Perspective," November, 2001, available online at www.tnsofres.com.

9. CyberAtlas, "E-Government May Not Mean Efficiency," undated document available online at www.cyberatlas.internet.com, November 30, 2001.

10. Mellor, Victoria Parr, and Hood, "Government Online."

11. Ibid.

12. Ibid.

13. Ibid.

14. Marcus Franda, *Launching into Cyberspace: Internet Development and Politics in Five World Regions*, Boulder: Lynne Rienner, 2002.

15. Ivan Katchanovski and Todd LaPorte, "CyberDemocracy or Potemkin E-Villages: Electronic Governments in OECD and Post-Communist Countries," *International Journal of Public Administration*, forthcoming.

16. Todd LaPorte, Chris Demchak, and C. Friis, "Webbing Governance: Global Trends Across National-Level Public Agencies," *Communications of the ACM*, Volume 44, 2001, pp. 63–67, and Todd LaPorte and Chris Demchak, "Hotlinked Governance: A Worldwide Assessment, 1997–2000," paper presented at the annual meeting of the American Political Science Association, San Francisco, Cal., August 30–September 2, 2001.

17. Pippa Norris, *Digital Divide?*

18. The 2001 study was undertaken in conjunction with the World Markets Research Centre consulting company of London.

19. The analysis was undertaken during summer 2001 at Brown University. Tabulation for this project was completed by Kim O'Keefe, Julia Fischer-Mackey, Sheryl Shapiro, Chris Walther, Shih-Chieh Su, Ebru Bekyel, and Mariam Ayad.

20. Council for Excellence in Government, "The New E-Government Equation," April, 2003. Unpublished survey report available online at www.excelgov.org.

21. For a detailed case study of Singapore, see Hairong Li, Benjamin Detenber, Wai Peng Lee, and Stella Chia, "E-Government in Singapore: Demographics, Usage Patterns, and Perceptions," unpublished research paper, 2003.

22. Data developed by Pippa Norris of the Kennedy School at Harvard University. They are available online at her Shared Global Database at http://ksghome .harvard.edu/~.pnorris.shorenstein.ksg/data.htm.

23. These data are from the Human Development Index developed by the United Nations Development Programme. They are available online at http://hrd.undp.org.

24. Data developed by Pippa Norris of the Kennedy School at Harvard University. They are available online at her Shared Global Database found at http://ksghome .harvard.edu/~.pnorris.shorenstein.ksg/data.htm.

25. These data are from the Human Development Index developed by the United Nations Development Programme. They are found online at http://hrd.undp.org.

26. Hargittai, "Weaving the Western Web," Kiiski and Pohjola, "Cross-Country Diffusion of the Internet," and Norris, *Digital Divide?*.

27. Data developed by Pippa Norris of the Kennedy School at Harvard University. They are available online at her Shared Global Database found at http://ksghome .harvard.edu/~.pnorris.shorenstein.ksg/data.htm.

28. These data are from the Human Development Index developed by the United Nations Development Programme. They are found online at http://hrd.undp.org.

29. Data developed by Pippa Norris of the Kennedy School at Harvard University. They are available online at her Shared Global Database found at http://ksghome .harvard.edu/~.pnorris.shorenstein.ksg/data.htm.

30. Also see United Nations, *World Public Sector Report 2003: E-Government at the Crossroads*, New York: United Nations, 2003.

31. Elaine Kamarck and Joseph Nye Jr., eds., *Democracy.com? Governance in a Networked World*, Hollis, N.H.: Hollis Publishing Company, 1999.

CHAPTER 10
DEMOCRITIZATION AND TECHNOLOGICAL CHANGE

1. National Performance Review, "From Red Tape to Results: Creating a Government That Works Better and Costs Less," Washington, D.C.: Government Printing Office, 1993.

2. David Garson, *Public Information Technology: Policy and Management Issues*, Harrisburg, Penn.: Idea Group Publishing, 2003.

3. Jon Gant and Diana Gant, "Web Portal Functionality and State Government E-Service," Hawaii International Information Systems—35 Conference Proceedings, 2002.

4. Karen Mossberger, Caroline Tolbert, and Mary Stansbury, *Virtual Inequality: Beyond the Digital Divide*, Washington, D.C.: Georgetown University Press, 2003.

5. Jerry Mechling, "Information Age Governance: Just the Start of Something Big?", in Elaine Kamarck and Joseph Nye Jr., eds., *Governance.com: Democracy in the Information Age*, Washington, D.C.: Brookings Institution Press, 2002, pp. 141–160.

6. Quoted in Elaine Kamarck and Joseph Nye Jr., eds., *Democracy.com? Governance in a Networked World*, Hollis, N.H.: Hollis Publishing Company, 1999.

7. Dennis Thompson, "James Madison on Cyberdemocracy," pp. 36–37, in Kamarck and Nye, eds., *Democracy.com? Governance in a Networked World*.

8. Theodore Lowi, "Four Systems of Policy, Politics and Choice," *Public Administration Review*, Volume 33, 1972, pp. 298–310.

9. Michael Hayes, "The Semi-Sovereign Pressure Groups," *Journal of Politics*, Volume 40, number 2, February 1978, pp. 134–161.

10. American Federation of Teachers Public Employees, "Digital Government and Technological Change: The Impact on Public Employees and Quality Public Services," 2002.

11. Ibid., pp. 1–2.

12. Ibid., p. 9.

13. Ibid., p. 12.

14. Lowi, "Four Systems of Policy, Politics and Choice."

15. Eszter Hargittai, "Radio's Lessons for the Internet," *Communications of the ACM*, Volume 41, number 3, January 2000, pp. 50–57.

16. Jane Fountain, *Building the Virtual State: Information Technology and Institutional Change*, Washington, D.C.: Brookings Institution Press, 2001.

17. *USA Today*, "Bush Proposed Free E-Filing, But Tax-Preparation Firms Hate That Idea," April 15, 2002, p. 8A.

18. Suzanne Garment, *Scandal: The Crisis of Mistrust in American Politics*, New York: Times Books, 1991.

19. Todd LaPorte and Chris Demchak, "Revolution or Evolution? Public Agencies, Networked Information Technologies and Democratic Values in the United States and Around the World," paper presented at the annual meeting of the American Political Science Association, San Francisco, Cal., August 30–September 2, 2001, and Todd LaPorte, Chris Demchak, Martin de Jong, and Christian Friis, "Democracy and Bureaucracy in the Age of the Web," paper presented at the International Political Science Association World Congress, Montreal, Quebec, Canada, August 2000.

20. Pew Research Center, Millennium Survey, released July 3, 1999.

21. Ronald Weber and Paul Brace, *American State and Local Politics*, New York: Chatham House, 1999.

22. Nelson Polsby, *Political Innovation in America*, New Haven: Yale University Press, 1984.

23. Richard Lourie, "Treasures of the Russian Evolution," *New York Times*, October 3, 2003, p. B31.

24. Adrian Johns, *The Nature of the Book: Print and Knowledge in the Making*, Chicago: University of Chicago Press, 1998.

25. Elizabeth Eisenstein, *The Printing Press as an Agent of Change: Communications and Cultural Transformations in Early Modern Europe*, New York: Cambridge University Press, 1979.

26. William Dill, "Growth of Newspapers in the United States," University of Kansas Bulletin, 1928, p. 53.

27. Simeon North, *History and Present Condition of the Newspaper and Periodical Press of the United States*, Washington, D.C.: Government Printing Office, 1884.

28. "Telephone: Fascinating Facts about the Invention of the Telephone," website www.ideafinder.com/history/inventions, April 9, 2002.

29. Darrell M. West, *The Rise and Fall of the Media Establishment*, Boston: Bedford/St. Martin's Press, 2001, pp. 28, 131.

30. Susan Douglas, *Inventing American Broadcasting 1899–1922*, Baltimore, Md.: Johns Hopkins University Press, 1987.

31. Darrell M. West, *The Rise and Fall of the Media Establishment*, Boston: Bedford/St. Martin's Press, 2001, p. 56.

32. Ibid., p. 59.

33. Theodore White, *The Making of the President*, 1960, New York: Atheneium, 1961.

34. Angus Campbell, Philip Converse, Warren Miller, and Donald Stokes, *The American Voter*, New York: John Wiley, 1960.

35. U.S. Bureau of the Census, *Statistical Abstract of the United States*, Washington, D.C., 1998, p. 572.

36. Internet Usage Data based on the Pew Internet National Survey, 2003. Data available online at www.pewinternet.org.

37. Nolan Bowie, "Voting, Campaigns, and Elections in the Future," pp. 120–144, in Anthony Corrado and Charles Firestone, eds., *Elections in Cyberspace: Toward a New Era in American Politics*, Queenstown, Md.: The Aspen Institute, 1996; Michael Margolis and David Resnick, *Politics as Usual: The Cyberspace "Revolution,"* Thousand Oaks, Cal.: Sage Publications, 2000; and Richard Davis, *The Web of Politics: The Internet's Impact on the American Political System*, New York: Oxford University Press, 1999.

38. Kenneth Kraemer, William Dutton, and Alana Northrop, *The Management of Information Systems*, New York: Columbia University Press, 1981; John King and Kenneth Kraemer, *The Dynamics of Computing*, New York: Columbia University Press, 1985; Kenneth Kraemer and John King, "Computing and Public Organizations," *Public Administration Review*, Volume 46, 1986, pp. 488–496, and Kenneth Kraemer and Jason Dedrick, "Computing in Public Organizations," *Public Administration Research and Theory*, Volume 7, number 1, 1997, pp. 89–112.

39. Jeffrey Abramson, Christopher Arterton, and Gary Orren, *The Electronic Commonwealth: The Impact of New Media Technologies on Democratic Politics*, New York: Basic Books, 1988.

40. Paul Abramson, *Political Attitudes in America*, San Francisco: Freeman, 1983, William Maddox and Stuart Lilie, *Beyond Liberal and Conservative*, Washington, D.C.: Cato Institute, 1984, and Joseph Nye Jr., Philip Zelikow, and David King, *Why People Don't Trust Government*, Cambridge, Mass.: Harvard University Press, 1997.

41. Helen Margetts, *Information Technology in Central Government*, London: Routledge, 1999.

42. Council for Excellence in Government, "E-Government: The Next American Revolution," April, 2000. Unpublished survey report available online at www.excelgov.org.

43. Judy Sarasohn, "Survey Finds Americans Split on 'E-Government,' " *Washington Post*, April 14, 2003, p. A14.

44. Dan Keating, "Pentagon Calls Off Voting by Internet," *Washington Post*, February 6, 2004, p. A12.

45. Michael Schrage, " 'ET' Phones Hill—With a TV Poll Of Ohio Viewers' Instant Views," *Washington Post*, March 10, 1983, p. E1.

46. Mossberger, Tolbert, and Stansbury, *Virtual Inequality*.

47. Federal CIO Council, "An Inventory of Federal e-Government Initiatives," January, 2001.

48. Roger Alcaly, *The New Economy: What It Is, How It Happened, and Why It is Likely to Last*, New York: Farrar, Straus and Giroux, 2003.

49. Mossberger, Tolbert, and Stansbury, *Virtual Inequality*.

50. Ben Shneiderman, "Universal Usability," *Communications of the ACM*, Volume 43, number 5, May, 2000, pp. 85–91.

51. "E-Government Act of 2002," text of legislation available online at http://thomas.loc.gov. Also see Rebecca Raney, "New Economy: In the Next Year, the Federal Government Will Move to Give the Public Easier Online Access to Data and Services," *New York Times*, December 23, 2002, p. C4, and Judi Hasson, "E-Gov Agenda Takes Shape," *Federal Computer Week*, December 2, 2002.

52. Cindy Skrzycki, "U.S. Opens Online Portal to Rulemaking," *Washington Post*, January 23, 2003, p. E1.

53. Amelia Gruber, "House E-Government Provision Falls $44M Short of Bush Request," GovExec.com, September 10, 2003.

54. Paul Volcker Commission Report, *Urgent Business for America: Revitalizing the Federal Government for the 21st Century*, released January 7, 2003.

Index

abortion issue, e-government information on, 42

Accenture, annual government website assessments of, 45–46

accessibility, 20; assessment of, 45; for disabled and foreign languages, 56–58; economic wealth and, 142; of global e-government, 151–152; inequity of, 182. *See also* disability accessibility; foreign language accessibility; World Wide Web Consortium (W3C), disability accessibility standards of

accessKansas Subscriber Services, 60

accountability, 47; in global e-government, 163; measures of, 10–11; mechanisms of, 31

accountability-enhancing features, 12, 108–109

administrators: of e-government, 181; evaluation of e-government by, 27t, 28t; personal involvement of, 27t; resistance of, 22; supporting e-government, 28–29; on top obstacles to e-government, 29t

Advanced Systems Resources, 138

advertisement, 58–59; on global e-government websites, 152–153, 163; on government websites, 34–35

Africa: online services of government sites in, 148; overall e-government performance in, 157; regional government websites of, 145

age groups, e-government usage by, 125, 126

agency missions: compatibility with, 31–32; integrated, 184; statements of, 46

agency websites: integrated information on, 180–181; nonresponsiveness of, 7; performance of, by type, 66–67, 68t; readability of, 54–55; stages of transformation of, 8–12

aggregate multivariate analysis, 19, 71–74, 140, 157, 159–161

Alaska: Department of Motor Vehicles of, 50; number of online services in, 74

Albany, New York, city website for, 40

Alvarez, Michael, 23

American Federation of Teachers Public Employees (AFT), 169–170

American Management System, 32

American Society for Public Administration analysis, 46

Antigua and Barbuda Portal/Tourism, 152

Arizona: e-government innovation of, 44–45; online voting innovations in, 1; tourism promotion in, 58–59

Arkansas Administrative Statewide Information Systems (AASIS), 41

Arkansas government websites: e-government controversies in, 137–138; readability of, 56; resources devoted to development of, 39–40, 41

ARPANET, 2–3, 178

Asia: online services of government sites in, 148; overall e-government performance in, 157; regional government websites of, 145

asynchronous communication technology, 2–3

audio clips, on global government sites, 148

Australia government websites: best practices of, 198–199; online services of, 150; overall performance of, 156; rating of, 45; security of, 151

Australian government websites: organization assessment of, 46

Austria government websites, 156

authoritarian governments: e-government performance in, 166–167; e-government sites of, 141

automobile, as secular change model, 8

Baheti, Arun, 138

Bangladesh, National Tourism Organization site of, 150

Beierle, Thomas, 11

Belgium, government websites of, 45

Belize Tourism, 152

Bell, Alexander Graham, 176

Bennett, James Gordon, 176

Benz, Karl, 8

Bhutan Portal, 153

billboard website stage, 8–9, 11t, 47

Bimber, Bruce, 5

Bobby service, automated online (Watchfire software), 151–152